STUDIES IN OPTICS

A. A. Michelson

With a Foreword by
Harvey B. Lemon

DOVER PUBLICATIONS, INC.
New York

Published in Canada by General Publishing Company, Ltd., 30 Lesmill Road, Don Mills, Toronto, Ontario.

Published in the United Kingdom by Constable and Company, Ltd., 3 The Lanchesters, 162–164 Fulham Palace Road, London W6 9ER.

Bibliographical Note

This Dover edition, first printed in 1995, is a slightly altered, unabridged republication of the work originally published by The University of Chicago Press, Chicago, in 1927. In the Dover edition Plate I, facing Page 168, appears in black and white in the text and has also been reproduced in color on the inside front cover.

Library of Congress Cataloging-in-Publication Data

Michelson, Albert Abraham, 1852–1931.
    Studies in optics / A.A. Michelson ; with a foreword by Harvey Brace Lemon.
        p.   cm.
    Includes index.
    ISBN 0-486-68700-7 (pbk.)
    1. Optics.
QC355.2.M53   1995
535—dc20   95-10180
                                                                              CIP

Manufactured in the United States of America
Dover Publications, Inc., 31 East 2nd Street, Mineola, N.Y. 11501

NOTE

In chapter xiii the various experiments which were carried out to detect the effects of the motion of the medium on the velocity of light are described. The last of these experiments is, of course, the well-known one of Michelson and Morley which established that the velocity of light is the same in all inertial frames of reference and provided the base for the special theory of relativity. In describing these ideas bearing on special relativity, Professor Michelson adopts a cautious attitude, sometimes giving the impression of skepticism. Such an attitude was justifiable at the time in view of the revolutionary character of the theory. However, at the present time the experimental basis for special relativity is so wide and the theoretical ramifications so many that there can no longer be any doubt about its validity.

In chapter xiv reference is also made to the "generalized theory of relativity." However, this theory represents a development along somewhat different lines and except in a very general way does not bear on the subject matter of these two chapters. The foundations of the general theory (unlike those of the special theory) are still in the process of change and evolution.

S. CHANDRASEKHAR

FOREWORD

ALBERT ABRAHAM MICHELSON, 1852–1931

In 1907 Albert Abraham Michelson became the first American citizen to win a Nobel Prize in science. The Swedish Academy gave recognition to Michelson's "optical instruments of precision and the spectroscopic and metrological investigations he had executed with them." A world figure, he managed to maintain such a quality of scientific work for more than a half century that his results in the most profoundly fundamental fields of study remain unquestioned as to fact. Michelson never set great store by his opinions, never mistook them for facts.

Michelson's entire scientific career, begun while a student at the United States Naval Academy and continued without pause until, in his seventy-ninth year, he suffered a cerebral hemorrhage that caused his death, is summed up for posterity in some 78 published papers. The first, consisting of one page, was printed in 1878 when he was 26 years of age and is entitled "On a Method of Measuring the Velocity of Light." The last, containing an introduction written by himself ten days before he lost consciousness, is on the same subject.

Light in some aspect or other was the topic of all but ten of these papers. Michelson's life work in this field lay chiefly along two avenues. The first was the accurate determination of the speed of light. In this field, neither the

A more extensive version of this Foreword appeared in *The American Physics Teacher*, Vol. IV, No. 1 (1936).

young man of 26 nor the old man of 79 has ever had a rival. It is remarkable testimony to the world-wide confidence in his ability, honesty, and judgment that during his life no one ever attempted to repeat his experiments or check his results on this subject excepting himself. Michelson's first interest in this problem was aroused when he was a student under Simon Newcomb, professor of physics at the Naval Academy. The latter had been carrying on an elaborate investigation supported by thousands of dollars of congressional appropriations and could well have been forgiven for incredulity and some irritation when his work was matched in accuracy by this stripling with an outfit built by himself and costing ten dollars. Newcomb writes under date of February 20, 1880, in an introduction to Michelson's paper officially required by naval etiquette, "To prevent a possible confusion of this determination of the velocity of light with another now in progress under official auspices, it may be stated that the credit and responsibility for the present paper rests with Master Michelson." Again in the preface to the official paper it is stated that "the degree of precision originally aimed at can be reached without any radical change in the instrument." The invitation was extended to cooperate with any other physicist who desired to use it (the official instrument) for further researches.

Michelson, however, had made a radical modification in the instrument. He shifted the rotating mirror from its classical position where it had been placed by Foucault, the first to use it, and where it was used by Newcomb, to a different location in the optical path. This shift enabled the light to be sent a far greater distance and back without excessive enfeeblement. It was this brilliant idea that made it possible for his crude apparatus to compete with the official one and that subsequently enabled Michelson, when he had greater resources at his command, to extend the path for many miles and so obtain precision undreamed of, and impossible by any other device that has been invented. Michelson's own last determination shows an accuracy of 3 parts in 10^6; that is to say, the 186,000-odd-mile journey made by light in one second is known to within half a mile.

Michelson entered on the second great avenue he was destined to pursue, the study of interference of light, also very early in life. At the age of 30 he published his sixth paper, an article of nine pages entitled "The Relative Motion of the Earth and the Luminiferous Ether." In this field he was destined to have many follow his lead, but still only two competitors ever came near to equaling his precision and none even remotely showed his wealth of originality and resourcefulness. This highway of discovery on which he entered was as fundamental as the other but, unlike it, led into regions which certainly even Michelson could never have foreseen.

One branch went into the microcosmos, the world of atoms hopelessly removed from direct human vision, and furnished us with as direct evidence as any that we have of the kinetic dance of molecules that we call temperature. His thirtieth paper, entitled "On the Broadening of Spectral Lines," gives little less than a direct demonstration of equipartition of energy in the thermal motion of atoms of no less than 17 different elements ranging in atomic weight from 1 to 210.

Another branch of his work on interference, *interferometry* as it has come to be called, led out into the macrocosmos of the stars. In 1890 his twentieth paper (21 pages, one of the longest he ever wrote), dealt with observations on a double star (Capella) too close for resolution by the best telescopes of the time. In the following year his twenty-first paper was on the "Measurements of Jupiter's Satellites by Interference"; this work was done at Lick Observatory. Thirty years elapsed before the astronomers reverted to his technic to measure the giant star Betelgeuse.

However, these majestic flights to outer space and then to inner space were but digressions, which never long diverted him from the end he had in view when starting on this avenue of study. This end was to discover, if possible, the absolute motion of the earth, as, trailing along with the rest of the solar family, it followed the sun's plunging course through space. The wave character of light had been established, by other experiments as well as his own on interference; and, in common with all other wave phenomena that are known, light was thought to require a medium for its propagation. This medium, the ether, was believed to be universal in extent and to afford a fixed frame in space to which all motions might be referred. Indeed, had it not been for the experiments which Michelson was about to undertake, we should probably today be living in a scientific world dominated by the same idea.

Shortly after leaving the Naval Academy and going to Case School of Applied Science at Cleveland, Michelson had repeated his experiment on the speed of light with better apparatus. He had also measured the speed of light in water and in carbon disulphide and found the values in these mediums to be in a complete agreement with classical theory which related the speed to the refractive index of the medium. What more natural, therefore, than that he should in addition send his beam of light upstream and back through the free ether of space and, by comparing the time required for this journey with that required for a similar beam to travel transversely across the current, thus determine the rate of flight of the earth through this fixed and all-pervading medium?

This experiment required an almost unbelievable delicacy of observation, and for the purpose he invented and built with funds provided by Alexander Graham Bell a device by means of which the light waves themselves were made to measure that excessively minute retardation that they would suffer in going upstream and back instead of across the current. This device was his "differential-refractometer," later called the *interferometer*. The theory of it was published in his eighth paper in 1882. It was not surprising that the first instrument built should have been found not to possess sufficient rigidity of structure. He gave it thorough trials at Berlin and at Potsdam. But the very manipulation of it produced considerably greater alterations in the light paths than those expected from its motion through the ether and this first result was negative. Consequently, during the next five years, in collaboration with Professor Morley, an apparatus was constructed of such stability that it was capable of measuring about 1/12 part of the displacement of the interference fringes expected as a result of the earth's motion through the ether. To the complete astonishment and mystification of the scientific world this refined experiment also yielded absolutely negative results. Again must we note at this point the universal confidence with which any experimental fact announced by Michelson was instantly accepted. Not for twenty years did anyone have the temerity to challenge his conclusion. Instead, the theoretical physicists accepted it and in one way or another began to rebuild their theories. Many of the proposals seemed bizarre at the time, and most of them were satisfactory only to their own authors.

It was not until 1905 when Albert Einstein, taking Michelson's experimental result as one fundamental postulate and the constancy of the speed of light as the other, bought out his paper on "The Special Theory of Relativity" that the world in general accepted his interpretation of these results as a cornerstone of its physical philosophies. While Michelson himself professed that subsequent developments of the general theory of relativity were beyond his understanding, he undoubtedly was of the opinion that Einstein's interpretation of his experiment was a correct one. Nevertheless, he took great pains to leave no stone unturned to test any alternative hypothesis which could be put to experimental test either before or after the appearance of the relativity theory. For example, when in 1897 it was proposed in explanation of his negative result that possibly the ether at the surface of the earth was dragged along with a speed so nearly equal to that of the earth that within the limits of accuracy of the experiment no change in motion could be detected, Michelson sent his two interfering pencils of light around a vertical rectangle of 50 by 200 ft., making the entire vertical circuit of Ryerson Laboratory at the University of Chicago. Over so long a path, irregularities in the air would completely destroy the requisite sensitivity, and consequently here for the first time he sent the beam in a vacuum. To this practice he later frequently made recourse, as in his last experiment on light's speed. This vertical interferometer showed a displacement of less than 1/20 of a fringe which indicated that, *if* the ether *were* dragged by the earth, this drag would extend out in space presumably to a distance of several thousand miles, a conclusion that common sense rejected as highly improbable.

Again, at the request of certain of the relativists, he spent many thousands of dollars building an apparatus along lines which he had himself proposed in 1904 and at that time rejected as showing positive evidence only of the earth's rotation in the ether but not of its translatory motion. Michelson engaged upon this expensive enterprise only out of deference to the desires of some advocates of the theory of relativity, whose mathematical arguments he modestly professed he was unable to refute. After his assent to the project had been obtained, he remarked, "Well gentlemen, we will undertake this, although my conviction is strong that we shall prove only that the earth rotates on its axis, a conclusion which I think we may be said to be sure of already." The result, published in collaboration with H. G. Gale in 1925, turned out precisely as he had prophesied. In this experiment both classical theory and the theory of relativity predicted the same results, a shift of 236/1000 of the distance between fringes. The experiment, performed on the prairies west of Chicago, showed a displacement of 230/1000, in very close agreement with the prediction. The rotation of the earth received another independent proof, the theory of relativity another verification. But neither fact had much significance. This is a clear example of Michelson's remarkable intuition and insight with respect to physical phenomena. One of his younger associates in the laboratory has written in this connection, "Michelson saw with intuition unsurpassed the logical relationship between phenomena, where others were forced to spend months in mathematical calculations to reach the desired results. His judgment in scientific matters was supreme."

A chapter of this puzzling question of ether drift was published in 1928 under the title, "A Conference on the Michelson and Morley Experiment." Michelson's part in this was his seventy-seventh contribution and the last of which he ever read the proof sheets. The conference was held at Pasadena in that year and resulted from the fact that, long before, Professor D. C. Miller, who had been associated with Morley at Cleveland after Michelson left to go to Clark University, had carried on at different times extending over a great number of years subsequent experiments on ether drift, first with Morley and then by himself. The tremendous importance of the problem fully justified Miller's long-continued attack, in which by the patient acquisition of one hundred thousand individual observations he hoped to achieve a precision correspondingly greater than Michelson's dozen odd of eariier years. The announcement in the scientific journals by Miller that a very minute residual of drift had been detected and that its orientation in space was approximately in the direction of flight of the solar system revived world-wide interest in the problem.

To this conference at Pasadena came Lorentz who with Fitzgerald had years before proposed a theory of a contraction of the instrument in the direction of motion, just sufficient to compensate the effect of ether drift. This Fitzgerald-Lorentz contraction had been the most widely accepted explanation of the negative result prior to the work of Einstein. Experimenters came likewise to this conference; notably Kennedy, who had designed an apparatus of great precision for repeating these experiments independently, and who had won expressions of enthusiastic admiration from Michelson for the beauty of his device. Piccard's results too were discussed: Piccard on one of his balloon ascensions had carried an interferometer 7000 ft. aloft and also had set one up on top of the Rigi. The result of all of these experiments, excepting Miller's, were absolutely negative, and the discussion as to the interpretation of Miller's experiments wound up in a state of utter confusion, as discussions so often do.

Other aspects of Michelson's activity which lay along this highway of research on interference should be mentioned. One of his earliest ideas was that of calibrating our standards of length in terms of the changeless wave lengths emitted by atomic systems. In 1887 he published in collaboration with Professor Morley a paper of three pages entitled, "On a Method of Making the Wave-Length of Sodium Light the Actual and Practical Standard of Length." His "Plea for Light Waves," an address in the following year before the physics section of the American Association for the Advancement of Science, expressed the same idea. This work came to its fruition six years later when, by invitation of the International Committee on Weights and Measures, Michelson set up an interferometer in Paris and found that the standard meter contained 1,553,163.5 waves of the red radiation of cadmium.

In his search for a sufficiently simple and homogeneous radiation to serve for so precise a purpose, Michelson had laid the foundations, in the whole field of spectroscopic measurements, of those technics that for a time enabled spectroscopists to boast that their work involved the most accurate measurements of the smallest quantities in the whole field of science. That light waves can be measured to a precision of about 1 part in 10^7 is due very largely to those trails blazed by Michelson as he digressed now and then from the main aim of his work on interference.

Explicitly and implicitly we have already referred to Michelson's willingness to co-operate with his colleagues. This was true of him even when these colleagues were specialists in altogether different fields. The most interesting example of his co-operative work is illustrated in his experiments on the determination of the rigidity of the earth. In the first decade of this century, geologists had begun to realize the significance of an idea originally proposed by Kelvin in 1863, that our earth, in spite of an interior temperature admittedly high enough to reduce almost any substance to the liquid state, nevertheless gives evidence from its dynamics as it spins around the sun of being approximately as rigid as steel. Experiments to verify this apparent paradox by measuring directly the earth's rigidity, by such notable men as C. C. and Horace Darwin, Schweydar, Heckel, and others, had given inconclusive results. These workers invariably had used delicate pendulum apparatus. The late Thomas Crowder Chamberlin, professor of geology at the University of Chicago, once asked Michelson if he could think of any way in which the problem might be attacked on the basis of a physical experiment. With characteristic reserve, Michelson replied that he would be glad to think about it and with characteristic dispatch a few days later informed his friend that he had a very simple method in mind by means of which he thought the desired result might be achieved. In none of Michelson's investigations, perhaps, did his simplicity and directness of attack show itself more clearly. The apparatus, a complete departure from what had been used formerly, was not only highly original, but simplicity itself. It consisted primarily of two 6-in. iron pipes 500 ft. long, buried in the ground and half filled with water. At either end of the pipes were pits where observations could be made. To avoid the tumult of a great city, the experiment was conducted on the campus of the Yerkes Observatory at Lake Geneva, Wisconsin. The earth's rigidity was measured by the tiny tides in these two artificial, sheltered, miniature seas. Observations made at first with simple microscopes proved too laborious, and the interferometer was again called into service. For such a problem, however, the interferometer would be far too sensitive. It had to be adulterated very generously to bring its record conveniently within the range of automatic recording by motion-picture cameras. Even then, it happened once or twice that the record for some unknown reason entirely disappeared, to return of itself again after an interval of several hours. Careful scrutiny at the time of these mysterious disappearances revealed nothing wrong with the arrangements. Subsequently it developed that there had been slight earthquakes in Japan and that a seismograph of surpassing delicacy unconsciously had been created. As might be expected, these results, published in their final form in collaboration with H. G. Gale in 1919, probably stand for all time as an enduring and exact record made by direct measurement not only of the earth's rigidity, but of its viscosity as well.

The most baffling problem of a purely mechanical sort that Michelson ever attempted and one that perhaps did most to wear down his vitality, especially in the later years of his life, was the ruling of diffraction gratings. Such devices in a crude form were first made by Norbert in 1851, but these consisted only of a few thousand lines. L. M. Rutherford, of New York, in 1868 ruled 2-in. surfaces having about 10,000 lines/in., but gratings were first brought to high and useful quality by Henry A. Rowland, of Johns Hopkins University, who succeeded in ruling surfaces 6 in. long containing a total of as many as 100,000 lines and having as the technical measure of their effective usefulness a resolving power of 15×10^4. Subsequently, after Rowland's death, his machine has produced gratings of 40×10^4 resolving power in the fourth order.

Since Rowland gratings were difficult to obtain because of their scarcity, Michelson in 1899 began the construction of a ruling engine of his own, thinking that with the experience that had been gained from his predecessors the task could be accomplished in about six months. After eight years of struggle, he produced a 6-in. grating containing 110,000 lines of a perfection measured by 60×10^4 approximately. In 1915 he achieved the production of both an 8- and a 10-in. grating, containing 117,000 lines, which are still among the most powerful instruments of diffraction that the world possesses.

No discussion of Michelson, however brief, can fail to comment on the man's artistic side. This artistic side crept unconsciously into his scientific bibliography in two papers little known. The first, entitled "Form Analysis," appeared in 1906. At this period, Michelson was strug-

gling with the ruling engine and doing some highly mathematical work to develop the theory of a reciprocal relation
in diffraction to which his intuition had led him and which
he had subjected to experimental verification. He begins
his paper on form analysis as follows:

As a recreation in the midst of more serious work, I have been
interested in the analysis of natural forms: and hoping that the
results of this somewhat desultory occupation be not deemed too
frivolous for so august an occasion [a meeting of the American Philosophical Society], I will venture to present some illustrations and
generalizations which have occurred to me. I recognize that the
subject is one whose adequate treatment would tax the best efforts
of one who combined the insight of the scientist, with the aesthetic
appreciation of the painter and the gift of language of the poet—and
certainly I am lacking in all three—but especially in the power of
adequate expression. I had hoped that my contribution would at
least have the merit of originality, but I find that many abler inves-
tigators have found a similar delight in this interesting field and
have expounded their ideas with a wealth of poetic imagery and of
exquisite illustration such as I cannot hope to emulate.

The paper consists of an elaborate and detailed classi-
fication of natural symmetrical forms that occur in strik-
ingly similar fashion in such widely dissimilar objects as
vegetables, protozoa, crystals, and even liquids. With a
vividness which belies his modest conviction that he lacks
in powers of adequate expression, he describes the forms
produced by drops of colored liquid falling into another
liquid of nearly the same density, and concludes as follows:

In designing for the sake of decoration, symmetrical forms are
everywhere manifest, and the perception of their mutual relations
is indispensable to the student of art. Occasionally, however, there
is in decoration a deliberate departure from symmetry, and such a
variation may greatly enhance the beauty and effectiveness of the
design. We tire of too great uniformity even of agreeable kinds, and

the element of variety is as important in art as an occasional discord
is in music—its purpose being to heighten the effect of the succeeding harmony.

One of the great disadvantages of the modern tendency to extreme specialization in research is the loss of companionship of the
sister sciences, with the attendant loss of perspective which a more
general survey of the whole field of science should furnish. Should
we not, then, utilize every opportunity which promises to further
their union?

The geologist, the chemist, the physicist, the mathematician,
may and occasionally do meet here on the common ground of
crystallography. By a comparatively slight extension, the "ground
forms" of organisms—as Haeckel terms them—may also be included with a corresponding extension of our society of sciences to
include zoology and botany.

Nay, Art will demand a chair at the banquet, and Music and
Poetry will also grace the feast.

The other paper is entitled "On the Metallic Coloring
in Birds and Insects." The text calls attention to the fact
that the rainbow and the halo are the only cases of prismatic colors in nature (those scattered by individual dew-
drops and ice crystals are presumably included). Almost
all other cases of color in nature are due to selective absorption, by pigments. However, two other physical methods produce color: diffraction and other types of interference, and reflection from surfaces of metals. Suspecting
that the so-called iridescent olors seen in hummingbirds,
certain kinds of butterflies, and a wide variety of beetles
and other insects were neither prismatic in character nor
due to pigments, Michelson set out to discover which of
the other two physical phenomena, interference or metallic reflection, might be their cause. His conclusion is that
with few exceptions these iridescent colors have precisely
the same physical attributes as colors reflected from thin

metallic films. One noteworthy exception, that of the diamond beetle, was found to have, in the iridescent spots upon its wing case, gratings as fine as 2000 lines/in. In the technic of making optical tests of high precision upon objects of such extreme minuteness, Michelson's experimental genius never shone with greater luster, and in the water-color paintings of his specimens which he provided for the illustrations in his *Studies in Optics* the artist in him found complete expression.

F. R. Moulton, in an appreciation of Michelson published in *Popular Astronomy* for June–July, 1931, clearly caught the spirit of his work:

He was unhurried and unfretful, he was never rushed by University duties, he never drove himself to complete a laborious task, he never feared that science, the University, or mankind was at a critical turning point. He never trembled on the brink of a great discovery. . . . There are doubtless many motives that inspire men to scientific achievements. If I have clearly caught the dominant note of his life, Michelson was moved greatly by the aesthetic enjoyment his work gave him. In everything he did, whether it was work or play, he was artistic. . . . He pursued his modest and serene way along the frontiers of science making new pathways and ascending to unattained heights as easily and as leisurely as though he were taking an evening stroll.

When asked by practical men of affairs for reasons which would justify the investment of large sums of money in researches in pure science, he was quite able to grasp their point of view and cite cogent reasons and examples whereby industry and humanity could be seen to have direct benefits from such work. But his own motive he expressed time and again to his associates in five short words, "It is such good fun."

HARVEY B. LEMON

PREFACE

The following pages are intended to give a résumé of my own investigations, somewhat after the manner of the treatment given in *Light Waves and Their Uses*, from which booklet many of the illustrations are taken, but with more attention to the theoretical side, and covering a number of investigations in which I have been occupied since the appearance of that publication. While much of the work is of a nature which involves a general acquaintance with the method of the calculus, it is hoped that a fair idea of the experimental method and results may be presented to those who may be interested in the subject, but who lack the necessary mathematical equipment.

In order to give the subject a semblance of continuity, it will be desirable to introduce considerable matter which will be found in any of the standard works on optics—but even here it may be of interest to present these investigations from my own point of view and to convey my own impressions in such a way as to emphasize the ideas of the founders of the science which have made the deepest impression on my own mind.

In the chapter on "Diffraction" I have selected freely from the scientific papers of Lord Rayleigh.

I take pleasure in acknowledging the assistance of Dr. H. G. Gale, and Dr. Henry Crew, in the reading of the proof sheets and of Dr. G. S. Monk, who re-read the proof and also furnished many of the illustrations.

A. A. MICHELSON

CONTENTS

I. THEORY AND APPLICATIONS OF INTERFERENCE OF
LIGHT-WAVES 1

II. INTERFERENCE OF LIGHT-WAVES 10

III. THE INTERFEROMETER 20

IV. LIGHT-WAVE ANALYSIS 34

V. MEASUREMENT OF STANDARD METER IN LIGHT-
WAVES 46

VI. DIFFRACTION 55

VII. TESTING OF OPTICAL SURFACES 73

VIII. DIFFRACTION GRATINGS 86

IX. THE RULING OF DIFFRACTION GRATINGS . . . 99

X. THE ECHELON GRATING 104

XI. APPLICATION OF INTERFERENCE TO ASTRONOMICAL
INVESTIGATION 111

XII. VELOCITY OF LIGHT 120

XIII. EFFECTS OF MOTION OF THE MEDIUM ON VELOCITY
OF LIGHT 139

XIV. RELATIVITY 156

XV. METALLIC COLORS IN BIRDS AND INSECTS . . . 167

INDEX 175

CHAPTER I

THEORY AND APPLICATIONS OF INTER-FERENCE OF LIGHT-WAVES

To account for the various phenomena of light, two theories have been proposed: the corpuscular and the undulatory. The former postulates that light consists of trains of corpuscles emitted by the luminous source, all traveling with the same enormously great speed, which either directly or by reflection enter the eye, where, impinging on the retina, they cause the sensation of vision. The undulatory theory, on the other hand, requires that light consist in the propagation of a series of undulations in a medium which permeates all space, and ascribes the propagation with the same high speed to the properties of this medium, the ether. Both theories give correct explanation of the ordinary phenomena of propagation, reflection and refraction; but whereas the undulatory theory satisfies all requirements without the necessity for accessory hypotheses, this is not the case with the corpuscular theory.

Analogy with known forms of wave motion lead to the same preference. Thus in the case of water waves—such as occur when the surface of a quiet pond is disturbed by the fall of a stone—the circular waves which travel outward from the center of disturbance can furnish all the necessary knowledge of the source. Thus the direction and distance may be determined by erecting normals at two points of the circular wave front—their intersection will give the direction and distance of the source. If this is a periodic disturbance, this character will be impressed on the waves, whence the frequency of the source and its character may be inferred; and, finally, the magnitude (intensity) of the source may be inferred from the height of the waves.

Exactly the same kind of information is conveyed to the ear by sound waves in the air; and the same is also true of the waves in the earth crust due to an earthquake. But this is also the only kind of information the eye receives and interprets; that is, concerning the direction, distance, intensity, and character of the source. Thus analogy would lead to this explanation even if other proof were wanting. But in one very important particular the two theories give diametrically opposite results, namely, that the speed of propagation of light-waves should be greater in air than in water, while that of light-corpuscles should be less.

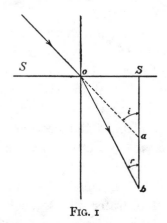

FIG. 1

This may be demonstrated as follows. According to the latter theory, a light-corpuscle, traveling in the direction of the arrow in Figure 1 and meeting the water surface *ss*, is attracted by the water particles; which by symmetry will affect only the normal component *sa* leaving the horizontal component *os* unaltered. Accordingly, the resultant direction is changed from *oa* to *ob;* and if these distances are proportional to the two velocities, respectively, the figure shows at once that $\sin i / \sin r = os/oa \div os/ob = ob/oa = v_2/v_1 = n$, the index of refraction. But as n is greater than unity, v_2 is greater than v_1; that is, the velocity in the second medium is greater than in the first.

The result in the undulatory theory is not quite so simply obtained. For this let us anticipate an important hypothesis, due to Huyghens, which states that if a wave front be given at any time, it may be found at any subsequent time by considering every point in the former as a separate source of disturbance, and constructing the spherical wavelets about these points with the radius corresponding to the distance the light would travel in the interval. The envelope of the wavelets so constructed will be the new wave surface. Suppose a plane wave ab, traveling in the direction of the arrows, meets the water surface ac.

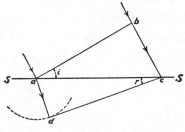

FIG. 2

While the ray at b with a velocity v_1 travels through a distance bc, the ray ad travels through a distance ad with the velocity v_2 and reaches the surface of a spherical wavelet of this radius. The new wave front will therefore be the plane surface passing through c and tangent to the sphere. The normal from a to the point of tangency will be the new direction. It follows readily that i and r are the angles of incidence and of refraction, respectively, and that

$$\sin i / \sin r = \frac{bc}{ac} \div \frac{ad}{ac} = \frac{bc}{ad} = \frac{v_1}{v_2} = n,$$

the index of refraction, which, being greater than unity, gives v_1 greater than v_2; that is, the velocity in the first medium is greater than in the second.

A celebrated experiment, which will be discussed in a future chapter, by Fizeau and Foucault, showed that the latter result is correct and the former is wrong. Apparently nothing further could be required to disprove the corpuscular theory; and were it not for the recent resuscitation of this theory in a modified and much more complicated form, it should hardly deserve mention in this connection.

The occasion for such a return to an exploded idea is a supposed difficulty brought to light by the theory of relativity, according to which, as many of its supporters maintain, not only is a medium unnecessary, but it is even inconsistent with its fundamental assumptions. But even if we admit the difficulty, it is doubtful if this is a valid reason for rejecting a theory which has yielded such splendid results. For it may well be that these difficulties —which, indeed, are not facts but deductions and interpretations of facts—may ultimately be modified in such a way as to explain the apparent inconsistency.

The particular form of wave motion required for the explanation of the phenomena of interference which will occupy our attention is not a matter of serious consequence. Even the direction of the vibrations is immaterial —save for a single investigation (on the cause of "metallic" colors in birds and insects), in which case the phenomena of polarization result from the transverse character of the vibrations; which, indeed, is a necessary consequence of the electromagnetic theory of light. Beautiful as it is, and powerful as a means of accounting for the various phenomena of light, it may be well to point out that it is in fact no explanation at all, in the sense of reducing the phenomena to more familiar types. Indeed, in this respect the elastic solid theory has all the advantage. For some sort of medium is required for the propagation of electromagnetic disturbances, and the properties of such a medium cannot be "explained" on any but mechanical concepts.

While it is true that the models hitherto devised for such explanation have not been entirely successful, they will be more acceptable to the average mind than none at all; and when it is considered what extraordinary properties this medium must possess—that while it offers no appreciable resistance to the motions of the planets in their orbits, it can transmit transverse vibrations, and with the inconceivably great velocity of 186,000 miles per second—such a medium must be constituted in a manner unfamiliar to our everyday experience. It is remarkable that our gross mechanical analogies have come so near to the truth. May it not be hoped that the very difficulties raised by the theory of relativity may prove a guide to such a structure and to such properties of a medium as will eliminate the difficulties themselves?

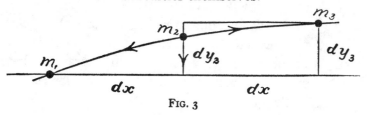

FIG. 3

To illustrate the deduction of the analytical expression showing the dependence of the velocity of propagation on the properties of the medium, consider the case of a stretched cord. Let the mass of the cord be concentrated at $m_1, m_2, m_3 \ldots$ at distances ds apart. The resultant force[1] on m_2 will be the resolved component of the tension T_1 in the direction of y, namely,

$$f = T_1 \frac{dy_2}{dx} - T_1 \frac{dy_3}{dx} \text{ or } m\frac{d^2y}{dt^2} = T_1 \frac{d^2y}{dx^2}dx ,$$

or since $m = \rho_1 dx (\rho_1 = \text{linear density})$,

$$\frac{d^2y}{dt^2} = \frac{T_1}{\rho_1}\frac{d^2y}{dx^2} .$$

The constant T_1/ρ_1 may be generalized to mean the ratio of restoring force to the inertia of the medium, and if designated by a^2 and the result extended to three dimensions, the vector s corresponding to the "displacement," the most general expression of the propagation of a wave motion, in a medium which has the same properties throughout and in all directions, is given by the formula

$$\frac{d^2s}{dt^2} = a^2\left(\frac{d^2s}{dx^2} + \frac{d^2s}{dy^2} + \frac{d^2s}{dz^2}\right) ,$$

in which s is the displacement, t the time, and x, y, z the co-ordinates of any point of the medium, supposed homogeneous and isotropic, and in which no other than "elastic" forces are acting. If the wave front—that is, the locus of all points in the same phase of vibration—be a plane normal to the direction of z, the expression reduces to the simpler form

$$\frac{d^2s}{dt^2} = a^2 \frac{d^2s}{dz^2} .$$

[1] The displacement is always assumed to be so small that the square may be neglected.

The solution of this differential equation is $s = f_1(z - at) + f_2(z + at)$, as is at once evident on substitution of this value in the original equation. f_1 and f_2 are any functions which are consistent with a physical existence.

Taking the function f_1 and giving to t an increment dt and to z the increment dz,

$$s_1 = f_1 \left[(z + dz) - a(t + dt) \right].$$

If now the increments are such that

$$\frac{dz}{dt} = a \text{ , then } s_1 = s \text{ ;}$$

that is, a is the constant velocity with which the form represented by f_1 is propagated in the direction of z. Similarly, f_2 is propagated in the opposite sense with the same velocity.

Let the function f take the simple periodic form

$$s = A \sin m(z - at)$$

or

$$s = A \sin (nt - mz)$$

or

$$s = A \sin 2\pi \left(\frac{t}{T} - \frac{z}{\lambda} + \psi \right),$$

in which s is the displacement, A the amplitude, T the period, λ the wave-length, and $\psi \cdot$the phase constant. This expression represents a homogeneous simple harmonic wave-train.

From the condition $dz = adt$, since λ is the distance through which the wave-train moves in the time T, it follows that $\lambda = aT$.

If several wave-trains are passing simultaneously through the medium, the resultant motion will be the vector sum of the components; and if s be the displacement of one of the components in a particular direction, and S the resultant in the same direction,

$$S = \Sigma s = \Sigma a \sin (nt - mx + \psi) .$$

In general, the resulting motion is not periodic. But if the periods are the same, the separate wave-trains differing only in amplitude and phase, then putting $nt - mx = \Theta$, which is the same for all,

or
$$\left.\begin{array}{l} S = \Sigma a \sin (\Theta + \psi) \\[2mm] S = \sin \Theta \Sigma a \cos \psi + \cos \Theta \Sigma a \sin \psi . \end{array}\right\} \tag{1}$$

But the resultant must have the same period as the constituents; so that if A is the resultant amplitude and a the resulting phase constant,

$$S = A \sin (\Theta + a) \tag{2}$$

whence
$$A^2 = \Sigma^2 a \cos \psi + \Sigma^2 a \sin \psi$$
$$\tan a = \frac{\Sigma a \sin \psi}{\Sigma a \cos \psi} .$$

If there are only two wave-trains,

$$A^2 = a_1^2 + a_2^2 + 2 a_1 a_2 \cos (\psi_1 - \psi_2)$$

$$\tan a = \frac{a_1 \sin \psi_1 + a_2 \sin \psi_2}{a_1 \cos \psi_1 + a_2 \cos \psi_2} .$$

If

$$\psi_1 - \psi_2 = 0 \qquad A_0 = a_1 + a_2$$

If

$$\psi_1 - \psi_2 = \pi \qquad A_\pi = a_1 - a_2$$

and if

$$a_1 = a_2$$
$$A_0 = 2a$$
$$A_\pi = 0 .$$

CHAPTER II

INTERFERENCE OF LIGHT-WAVES

When two similar wave-trains traveling in approximately the same direction are superposed, the resulting motion may be greater or less than that of the components, according to the difference of phase of the components. Thus if the two wave-trains are equal simple harmonic and meet in the same phase, the amplitude will be doubled and the intensity quadrupled. If, however, the phases be opposite, the resulting amplitude (and intensity) will be zero. In this case the two wave-trains are said to "interfere," and the resulting phenomenon is known as "interference." The term is not very well chosen, for in fact each train produces its own effect quite independently of the other, but it has been in use so long that it would not seem wise to alter it. Many illustrations might be given of such cases, but the following will suffice.

If two exactly similar tuning forks mounted on their resonators be sounded simultaneously, the resulting sound will not differ from that of either fork sounding alone, except that the resulting intensity will be different. But if one of the forks be loaded with a pellet of wax, its rate of vibration will be lowered, so that the two are no longer in tune. Supposing that the phase is the same at starting (and the resulting intensity great), in a short time the phases will differ and ultimately become opposite, when the resulting sound will be much weaker and may cease entirely for a brief instant. The difference of phase will thenceforward produce the regular alternations of sound and silence known as "beats." In this case the intensity is a periodic function of time.

If the two forks are exactly alike and are kept in the same phase of vibration (e.g., by an electrical connection)

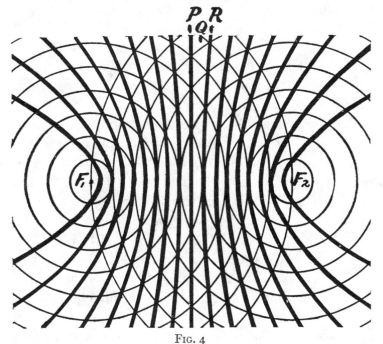

FIG. 4

a point at P (Fig. 4) equidistant from F_1 and F_2 will show a maximum of sound, whereas a point Q, whose distance is a half-wave farther from F_2 than from F_1, will always be the location of a point where the phases are exactly opposite, hence at this point there will be silence. At R, where the difference in path is a whole wave, the two sounds will again be in the same phase and here there will again be a maximum, and so on. In this illustration the intensity is a periodic function of the distance along any line parallel to F_1F_2.

If now light is a wave motion, the intensity being proportional to the square of the amplitude, then under proper conditions light added to light may produce darkness. It was Newton who first studied this phenomenon by means of the celebrated experiment of "Newton's rings." Doubtless the similar phenomena of the beautiful colors in soap bubbles had been observed centuries before, but without any idea of their cause; and it required the genius of a Newton to give the phenomenon a form which permits of an exact measurement.

It is true that Newton's explanation of the "colors of thin films" is no longer accepted (Newton opposed the wave theory of light and held to the corpuscular theory which required accessory hypotheses to account for the phenomenon); but the fact remains that he did actually measure the quantity which is now designated as the wave-length, and showed that every spectrum color is characterized by a definite wave-length.

The theory of the colors of thin films will be treated presently in some detail, but it may be noted here that it is the result of the superposition of the two beams of light, the one reflected from the first surface of the film (of water in the case of the soap bubble, and of air between two glass surfaces in the experiment of Newton's rings), the other that from the second surface; the "interference" between them produces the alternations of bright and dark bands if the light be of one pure color. If red light is employed, it will be observed that the circular bands in Newton's experiment are larger than if the light were blue; consequently, if white light is employed

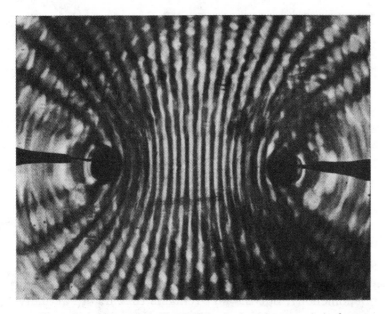

FIG. 5.—A photograph illustrating the interference of ripples produced by needle points through which an alternating current was passed and touching a water surface.

FIG. 6.—The shading representing interference in the space to the left of the second screen is, of course, only diagrammatic.

(which, as Newton showed, is made up of all the spectral colors), the result is the beautiful succession of colors so characteristic of the phenomenon.

Attention may be called at this point to an apparent difficulty in the explanation. The theory would seem to require that where the two surfaces are in contact the difference of path being zero, there should be a maximum of light, whereas experiment shows a minimum. The difficulty is at once removed when it is shown (as is also required by theory) that the two reflections do not take place under similar circumstances. One occurs at the rarer medium (air) while the other takes place at the glass surface, which necessitates a change of phase of just a half-wave. This must be added to the measured difference in path of one or the other of the two interfering pencils of light.

Now while it would appear that the complete explanation of the phenomenon is furnished by the undulatory theory, the crucial test could not be applied because of the impossibility of separating the two interfering pencils of light. This was accomplished by Thomas Young as shown in Figure 6 (p. 12). A minute aperture S on which sunlight is concentrated by a lens acts as a source of light which is intercepted by a screen having two small apertures a and b, about a millimeter apart. The light passing through reaches a screen a meter distant. At p where the two paths ap, bp, are equal, there will be a maximum of illumination, while at another point q where the difference in path $aq - bq$ is one half-wave, and where the phases are exactly opposite, the two pencils neutralize each other and there is a minimum of light. At a distance twice as great from p, the phases will again be alike producing a second maximum and so on. The strict analogy with the tuning-fork experiment previously described is manifest; and the same explanation applies in both cases.

The chief point of difference is that of wave-length, which necessitates a different order of distance between the interfering sources. If b is the width of the interference bands (distance from maximum to maximum), D the distance bp between the screens, d the distance between the apertures, and λ the length of the light-wave for the color of the light employed, then

$$b = \frac{D}{d}\lambda \ .$$

For white light the interference bands will be colored, the succession of colors being essentially the same as in Newton's rings.

Young found for the wave-lengths of the different colors the same result as Newton. In this experiment the two interfering pencils are separated so that they may be modified independently. Thus, if one of the pencils is suppressed, the interference bands vanish, and it must follow that here we have clear proof that (at the dark bands) light added to light produces darkness—a result entirely consistent with the wave theory, but very difficult to explain on any other hypothesis.

It will be noticed, however, that the pencils of light have traversed very small apertures, and that they are bent or diffracted from their original direction, and it might be objected that the results obtained were due in part or wholly to the effect of the edges of the screen on the light-pencils which have passed through the narrow apertures. To avoid this objection, Fresnel devised the following modification of Young's experiment.

While the method of the Fresnel mirrors as a means of illustrating the phenomenon of interference is superseded by more modern devices, this celebrated experiment contains many elements of interest both historically and from an educational standpoint.

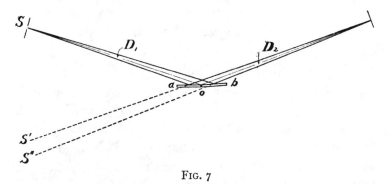

FIG. 7

Two glass plates, oa and ob, are inclined at a small angle (a few minutes of arc), and the light from a bright but very narrow source S is reflected by the mirrors to a screen where the resulting interference bands may be observed directly, or by an eyepiece (in which case the screen is unnecessary). The resulting illumination is the same as it would be if the original source and the mirrors were removed and replaced by the two virtual images of the source at S_1 and S_2.

If a_1 is the angle between the mirrors, D_1 the distance from their intersection to the source, and D_2 the distance to the screen, the width of the interference bands b for light of wave-length λ will be

$$b = \lambda \frac{D_1 + D_2}{2D_1 a_1}.$$

If the distances are equal,

$$b = \frac{\lambda}{a_1}.$$

It is important that the line of intersection of the mirrors be very near to the adjacent edges, otherwise the difference in path will be increased by $\dfrac{2h \sin \theta}{\lambda}$, in which h is the "offset" at the junction and θ the grazing angle. An error in this adjustment of 0.1 mm may amount to forty light-waves unless θ is small. It is desirable, therefore, to make the grazing angle very small.

If the source be a narrow slit (which gives a much more intense illumination than would a pinhole), it will also be quite important that the slit and the intersection of the two mirrors be in the same plane. It is to be noted, however, that owing to the obliquity of the reflection from the Fresnel mirrors[1] a considerable proportion of the light is also diffracted, and the objection raised against the experiment of Young still holds.

In another form of the Fresnel experiment the mirrors are replaced by a "bi-prism" whose angle is very slightly less than 180°. Here also the source s and the bi-prism produce the same effect on a screen s as would result from the two virtual images s_1 and s_2; whence it follows that the distance between the interference bands will be

$$b = \frac{\lambda}{2(\mu - 1)a_1} \cdot \frac{D_1 + D_2}{D_1},$$

[1] By careful adjustment of the intersection of the mirrors and by use of a rather high magnification, the obliquity of the reflected rays may be as small as desired, and thus the proportion of diffraction made insignificant.

in which a_r is the acute angle of the prism and μ the index of refraction. But as μ varies with λ, the resulting interference phenomenon is complicated by dispersion.

Before considering other forms of interference apparatus, it will be worth while to examine the principles which they have in common.

1. The first and most important of these is that the two interfering pencils must have a constant phase relation (or at least one whose variations are slow and continuous). This can only be realized in the case of light-vibrations if the two pencils originate in the same source. Thus it is altogether impossible to observe interference with two candles as sources; for the vibrations of the individual electrons, being practically independent, give a resulting illumination which amounts to an integration due to wave-trains whose phase, amplitude, and orientation vary many millions of times per second. This condition is therefore indispensable; the others which follow are convenient for facilitating the observation.

2. If the source is not homogeneous, that is, if it is made up of a mixture of colors (as in the case of white light), and hence of different wave-lengths, the resulting interference bands will have widths and positions depending on the separate wave-lengths and will coincide exactly only for the case of exact equality of the paths of the two interfering pencils, and approximately where this path difference amounts to a very few thousandths of a millimeter. Hence it will be important to make the two paths as nearly equal as possible,[1] unless the resulting light be examined spectroscopically, in which case the

[1] This condition would be violated, for instance, if the offset of the Fresnel mirrors is a hundredth of a millimeter or more.

interference phenomena are brought in evidence by the "channeling" of the spectrum; that is, the appearance of dark bands across the spectrum, the more numerous as the difference of path increases.

3. A third condition is that the direction of the two pencils should be nearly the same; otherwise, the interference fringes will be too narrow to distinguish. In fact, if a is the angle between the two pencils (and consequently between their wave-fronts), λ the wave-length or distance between two successive wave-fronts in the same phase, and b the distance between the bands, then

$$b = \frac{\lambda}{a} \, .$$

(This applies to all cases, as may readily be verified in those already considered. Thus, in the case of the Fresnel mirrors,

$$b = \frac{D_1 + D_2}{2 D_1 a_1} \lambda \, ,$$

where a_1 is the angle between the mirrors. But

$$2 a_1 = \frac{D_1 + D_2}{D_1} a \, ,$$

so that

$$b = \frac{\lambda}{a} \, ,$$

as required.)

If the difference in direction be, say ten minutes of arc, the breadth of the fringes will be only 340 wave-lengths, or about two-tenths of a millimeter.

There is still another condition which, however, applies only to apparatus like that of Young or of Fresnel, namely, that the source must be of very small dimensions (pinhole or narrow slit), otherwise the different systems of fringes are not superposed in the same location, thus masking all evidence of interference. There are other forms of apparatus, however, in which this condition is unnecessary, such as the interferential refractometer of

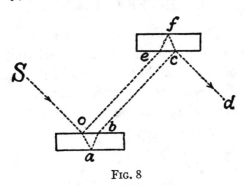

FIG. 8

Jamin. This consists essentially of two plane-parallel plates of exactly equal thickness and approximately parallel. One of the two pencils into which the light from an extended source at S is divided at the first surface follows the path *oabcd;* the other, the path *oefcd.* The two paths being nearly equal and the two pencils meeting along *cd* at a very small angle, interference fringes are readily observed either with the unaided eye or by means of an observing telescope.

CHAPTER III

THE INTERFEROMETER

The preceding methods of producing interference do not permit a wide separation of the interfering pencils of light, nor of considerable variations in the difference in path—conditions of importance in a number of investigations, among which may be mentioned the effects of temperature and pressure, of electric or magnetic fields, and of the effect of the motion of the medium upon the propagation of light.

It was for the investigation of this last-mentioned problem that the following arrangement was devised. Light from the source S, which may be an extended luminous surface (a candle, lamp, or a lens with an arc light at the focus, etc.), falls on the surface[1] of a plane-parallel glass plate a, which separates it into two "coherent" pencils, the one transmitted and the other reflected normally from the plane mirror c, and the latter from b, both returning to the separating surface a, whence they both proceed in the direction ae where the resulting interference fringes may be projected on a screen or observed by the eye, with or without an observing telescope.

It was this simple form to which the designation "interferometer" was originally applied, and which in

[1] This surface has a film of silver, or better, of platinum, of such thickness that the reflected and transmitted pencils are of approximately equal intensities.

slightly modified form was used in the experiment undertaken in 1880, and subsequently in collaboration with
Professor Morley, and which will be described in detail in
the chapter dealing with
the effect of motion of the
medium on the velocity
of light. Meanwhile, it
may be of interest to indicate some of the modifications which have been utilized for this and other investigations, and to point
out the analogies with
other optical instruments.

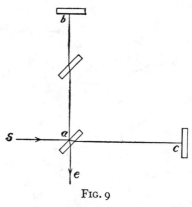

FIG. 9

For optical measurements two sorts of instruments are
in use: to wit, lenses (or mirrors) and prisms (or gratings).
The interferometer is sometimes added as a third; but

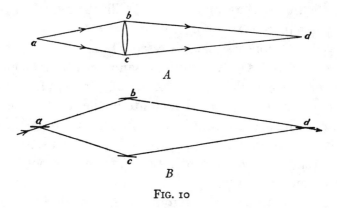

FIG. 10

since all optical phenomena depend essentially upon interference in its most general conception, there is no essential difference between them. The close analogy is illustrated in Figure 10. Thus in Figure 10A the image of

a source (a slit a, or a fine line ruled by a diamond on a smooth glass or metal surface) is formed at d (the result of the "combination" of all the rays which fall on the lens bc), where it may be ob-

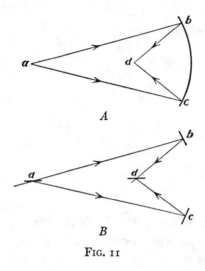

served, as in the telescope or the microscope, by an eyepiece. In Figure 10B the source is replaced by the surface a, whence two of the pencils (one transmitted and the other reflected) are bent by the prisms (or mirrors) at b and c so that they meet at the surface d, proceeding thence to the eye or the observing telescope.

A

B

Fig. 11

In Figure 11 the same analogy is illustrated when the lens is replaced by a mirror.

Thus it appears that the essential difference between lenses or mirrors, on the one hand, and the interferometer, on the other, is that in the former all the rays from the source which fall on the lens unite in the focal plane to form an image; whereas in the interferometer there are only two interfering pencils. As regards the use of the two classes of instruments, this difference is not very important; and indeed, it will be shown that for accurate measurements the interferometer has a decided advantage. In the illustrations just given, the microscope or telescope, and the analogous forms of interferometers, may be applied to the measurement of distances or of angles.

But prisms and gratings are employed in what seems at first sight to involve different principles, and for a different purpose, namely, the analysis of light into its

component constituents. The analogy still holds, however, as is shown in Figure 12A and B. Thus in Figure 12A, a represents the slit source of light and bc a grating which diffracts the light back to a (part being thrown one side by the plane-parallel plate p for observation or for photography), while in Figure 12B and 12C, the interferometer shows a similar light-path, but only for the two limiting pencils of light.

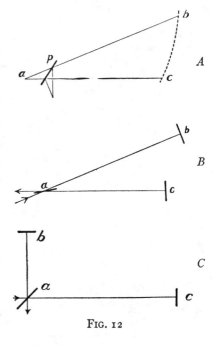

FIG. 12

If in this arrangement one of the mirrors, say c, is movable, and the incident light monochromatic, of wavelength λ, and if n is the number of maxima (or minima) corresponding to d, the measured difference in path, then the wave-length is given by $\lambda = d/n$; and, as will be shown in a future chapter, this can be measured with far greater accuracy than is possible by the use of prisms or gratings.

Figures 12C, 13C, 14B, 15C, and 16B represent the principal forms of interferometer together with their analogues for the special investigations to which they are adapted.

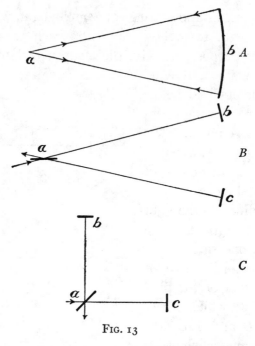

FIG. 13

The form of interferometer which has proved most generally useful is that represented schematically in

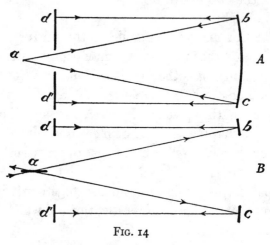

FIG. 14

Figure 13C, and in greater detail in Figures 17 and 18, and in the photograph shown in Figure 19. Light from the source (Fig. 17) falls on the lightly silvered surface of the plane-parallel plate A, where it divides, part being reflected to the plane-silvered surface C, whence it is re-

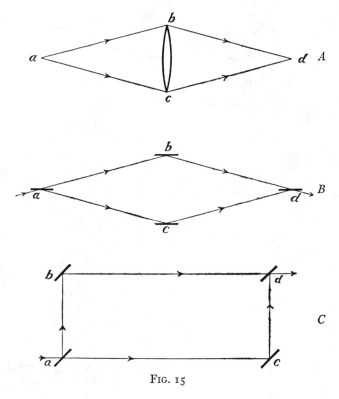

FIG. 15

turned to A, which transmits it to the observing telescope; the other part being transmitted to the plane mirror D, whence it is returned to A, being then reflected to the observing telescope, where it interferes with the first pencil. In white light, the optical difference in path must be very small if interference bands are to appear.

But one of the light-pencils has had to pass twice through the plane-parallel plate *A;* the compensating plate *B,* of

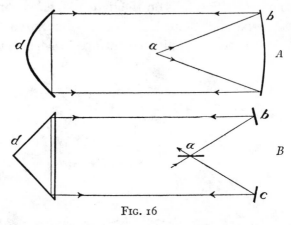

FIG. 16

exactly the same thickness, and placed at the same angle, is therefore interposed in the path of the second pencil.

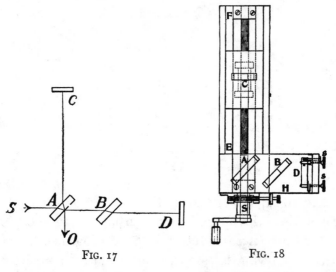

FIG. 17   FIG. 18

A heavy casting serves to support the optical parts, and the carriage holding the movable mirror *C* moves on very accurately ground ways. The motion is communicated by means of a screw provided with a worm wheel and a divided circle so that the motion of the carriage may be accurately measured. The stationary mirror *D* is provided with screws for adjustments about vertical and horizontal axes. The compensating plate *B*

FIG. 19

is held by a vertical steel rod, the torsion of which produces any required small alteration in the path. All of the optical surfaces are very accurately plane, the errors being of the order of a twentieth of a light-wave, or less.

The adjustment of the instrument is effected as follows. The distances of the mirrors *C* and *D* from the half-silvered surface of *A* are made approximately equal (say, to within a millimeter), and an approximately homogeneous source of light (sodium flame, or better, a Cooper-Hewitt mercury arc) is placed in front of A. The two images of a needle point are then brought into coincidence by the adjusting screws of the mirror D, when the interference fringes should appear.[1] They are usually narrow, curved, and not very distinct; but by slowly altering the adjustment of the mirror D they may be given any suitable width, and by diminishing the path difference by turning the screw S the fringes become more distinct. As the path difference approaches zero, the change of inclination of the fringes accompanying a change in position of the eye diminishes; and when this change vanishes, the (colored) fringes in white light appear, or may be found in a few turns of the worm wheel which gives the slow motion to the screw S.

THEORY OF THE INTERFEROMETER

In the arrangement of Figure 20, representing schematically the interferometer in its simplest form, it is evi-

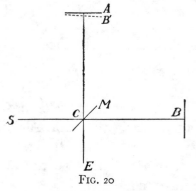

FIG. 20

dent that to the observer at E the effect of the combination is the same as that which would be produced by the surfaces A and B', the latter being the image of B in the mirror M. Let the pair of surfaces M_1M_2, Figure 21, replace the interferometer, and consider the effect of the reflection of light from a

---

[1] The clearness of the fringes at this stage may be considerably increased by the use of a small aperture held before the eye.

point S' to another point P. Let t be the distance between the surfaces at the points of incidence, and let ω be the angle of incidence, which may be considered the same for both mirrors, the angle ϕ between them being supposed very small (of the order of a second of arc, or less). The difference in path of the two interfering pencils will be

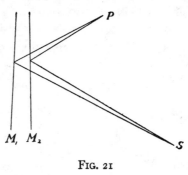

FIG. 21

$$\Delta = 2t \cos \omega .$$

Let the intersection of the mirrors be vertical, and let i be the angle between the horizontal projection of the incident ray and the normal; let θ be the angle which the projection of the ray on the vertical plane containing the normal makes with the normal. If P be the distance from the mirrors to the point where the interference fringes are observed, and if t_0 be the distance between the mirrors at the foot of the perpendicular, then

$$t = t_0 + P \tan i \tan \phi ,$$

or, since ϕ is very small, and ω is also small,

$$\Delta = 2(t_0 + Pi\phi) \cos \omega .$$

With an unlimited aperture the corresponding phase difference, $2\pi \dfrac{\Delta}{\lambda}$, may have so large a range of values that all traces of interference vanish. If, however, the cone of rays is limited (by the pupil of the eye, or by a diaphragm in front of the observing telescope), the range may be small enough to show the phenomenon of interference.

Let it be required to find the distance P at which the fringes will be most clearly visible. This must occur when the change in Δ is least for a given change in ω.

Putting $\dfrac{d\Delta}{d\omega}=0$, and since ω, i, and ϕ are all small,

$$P=\frac{it_0}{\phi}.$$

It follows that, for an unlimited beam, the different parts of the interference pattern are not simultaneously in focus, except, first, if $t_0=0$, when $P=0$ (fringes localized at surfaces of mirrors MM'; second, if $\phi=0$, when $P=\infty$ (fringes localized at infinity).

If Δ_0 is the phase difference for $\omega=0$, and if $\Delta_0-\Delta$ $=n\lambda$,

$$2t_0(1-\cos\omega)-2Pi\phi\cos\omega=n\lambda,$$

or for small ω,

$$t_0\omega^2-2Pi\phi=n\lambda,$$

or, if $\rho^2=\dfrac{n\lambda}{t_0}$, and if $h=\dfrac{P\phi}{t_0}$,

$$i^2+\theta^2=\rho^2+2hi. \tag{1}$$

This represents a circle whose angular semidiameter is

$$\sqrt{\frac{n\lambda}{t_0}+\frac{P^2\phi^2}{t_0^2}},$$

and whose center is displaced through an angle $h=\dfrac{P\phi}{t_0}$.

If t_0 is small, equation (1) reduces to

$$n\lambda = -2P\phi i = 2\phi x ,$$

or, disregarding sign,

$$x = \frac{n\lambda}{2\phi} ,$$

giving straight fringes parallel to the intersection of the mirrors. The distance between the fringes is

$$b = \frac{\lambda}{2\phi} .$$

VISIBILITY OF FRINGES

Let the visibility be defined by

$$V = \frac{I_1 - I_2}{I_1 + I_2}$$

in which I_1 and I_2 are the maximum and minimum intensities respectively.

$$I = \int_{i_1}^{i_2} \cos^2 \tfrac{1}{2} k\Delta di .$$

(The phase change for θ is of second order.)

The angle ω is always small so that $\Delta \doteq 2(t_0 + Pi\phi)$ and

$$I = i_2 - i_1 + \tfrac{1}{2} \int_{i_1}^{i_2} \cos 2k(t_0 + Pi\phi)di$$

$$I = (i_2 - i_1) + \frac{\sin kP\phi(i_2 - i_1) \cos k\Delta}{2kP\phi}$$

$$\therefore \quad V = \frac{\sin kP\phi (i_2 - i_1)}{kP\phi (i_2 - i_1)} \quad (k = 2\pi/\lambda) ,$$

This expression gives for $t_0 = 0$ (and therefore $P = 0$) the value $V = 1$, which shows that for moderate apertures the visibility is independent of the aperture. This is also true if $\phi = 0$. An idea of the order of magnitude of the angular aperture permissible may be obtained from elementary considerations as follows.

If the difference in phase between the central and the marginal rays be λ, the discordance will be considerable. In the first case, if β = angular aperture of the objective

$$\Delta = 2t(1 - \cos \beta/2) = \lambda$$

whence

$$\beta^2 = \frac{\lambda}{t} \, .$$

Thus, if $t = 25\lambda$, $\beta = 1:5$; that is, an angular aperture of the lens should be less than $1/5$.

In the second case, if A is the diameter of the objective $\lambda = 2\phi A$ whence A should be less than $\lambda/2\phi$.

INTERFERENCE FRINGES IN THE RECTANGULAR INTERFEROMETER

The interferometer is made up of two plane-parallel plates and two plane mirrors, placed at the corners of a rectangle, the mirrors and plates all being adjusted very accurately parallel, and at 45° with the rectangular path of the pencil of light from a broad source of approximately monochromatic light (a Cooper-Hewitt lamp). The fringes are viewed by a telescope. This arrangement is equivalent to an air plate of thickness t on which the light is incident at an angle of approximately 45°.

Let ϕ = inclination of surfaces of air plate; ψ = angle of their intersection with the horizontal; t = thickness of plate at $i=0$ and $\theta=0$; $(45+i)$ and θ, respectively, the horizontal and vertical projections of angle of incidence; $P=$ distance from the surface at which fringes appear for $i=0$; $x_1 y_1 =$ co-ordinates (parallel to surface) in "focal" plane. Then if $p = \bar{V}2\,P\phi\,\sin\,\psi - t$; $q = P\phi\,\cos\,\psi$; $r = x_1\,\sin\,\psi + y_1\,\cos\,\psi$; $\Delta =$ difference in path; the expression for the position of the interference fringes will be (omitting small quantities of second order):

$$pi + q\theta = \bar{V}2\Delta - r \ .$$

The isochromatic lines are straight, and inclined at an angle whose tangent is p/q. If α and β are the horizontal and the vertical angular apertures of the cone of rays entering the telescope, the visibility of the fringes is expressed by

$$V = \frac{\sin\cdot\omega}{\omega} \cdot \frac{\sin\,\chi}{\chi} \ ; \qquad \omega = \frac{2\pi}{\lambda}\,p\alpha\,, \qquad \chi = \frac{2\pi}{\lambda}\,q\beta\ .$$

It follows that V is fair only when $p\alpha$ and $q\beta$ are small; that is (for given apertures), when

$$P\phi\,\cos\,\psi \doteq 0 \ ; \qquad \bar{V}2\,P\phi\,\sin\,\psi \doteq t \ .$$

Hence the fringes are distinct only when $\psi = 90°$ and at a distance

$$P = t/\bar{V}2\,\phi \ .$$

If ψ is not $90°$, the fringes may be still readily visible if the aperture is small. The inclination is still $\tan\,\gamma = p/q$, and the breadth β is given by

$$4\frac{\lambda^2}{\beta^2} = (\bar{V}2\,P\phi\,\sin\,\psi - t)^2 + (P\psi\,\cos\,\psi)^2 \ .$$

CHAPTER IV

LIGHT-WAVE ANALYSIS

As has been indicated in the general discussion of the various forms of interferometer, these may be classified according to the uses to which they are applied. Thus, in the form C, Figure 13, in which the two paths are approximately equal, a motion of one of the mirrors of a hundred thousandth of an inch would correspond to a shift of one fringe, so that such an arrangement may be used to replace the microscope in measuring extremely small displacements. Similarly, the form of interferometer shown in B, Figure 14, is adapted for measuring extremely small angles, and may thus replace the telescope. It will be shown in the chapter on "Accuracy of Optical Measurements" that measurements made by the interferometer are from twenty to fifty times as accurate as the corresponding measurements by microscope or telescope.

The spectroscopic analogue is not quite so evident, though no less exact. If one of the mirrors of the interferometer be moved through a considerable distance, using a homogeneous source of light, counting the number of fringes which pass during the motion, the length of the light-wave will be twice this distance divided by the number of fringes. As shown in a future chapter, the accuracy of such a determination of wave-length is far greater than is obtained by the spectroscopic measurements.

If, however, the source is not homogeneous, the clearness or visibility of the fringes will diminish as the difference in path of the two interfering pencils increases, and in general will vary in a manner depending on the nature of the source. Thus, if the source is light from vapor of sodium, the fringes diminish in clearness until the difference in path contains n wave-lengths of λ_1 and $n+\frac{1}{2}$ wavelengths of λ_2. From this point they begin to grow clearer, until at twice this distance they are nearly as distinct as at zero. The visibility curve is given in Figure 22.[1]

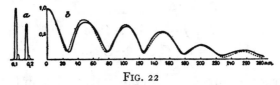

FIG. 22

Reciprocally, if such a visibility curve is observed, it follows that the source is a doublet. Similar considerations show that there is a close relation between the distribution of light in the source as a function of the wavelength and the corresponding visibility curve. The general analytical expression for this relation may be obtained as follows.

The intensity of the light from two interfering pencils of the same wave-length, λ, is

$$i = a_1^2 + a_2^2 + 2a_1a_2 \cos 2\pi \frac{D}{\lambda},$$

in which a_1 and a_2 are the amplitudes of the two wave-trains, and D the difference in path. If the two amplitudes are equal,

$$i = 2a^2\left(1 + \cos 2\pi \frac{D}{\lambda}\right).$$

[1] In Figures 22 and 24–29 inclusive, the dotted curve is the graph of the theoretical formula; the full curve represents the results of observation.

If, however, the light is not monochromatic, the total intensity will be

$$I = \int i \, d\lambda \,,$$

or,[1] putting $1/\lambda = n+x$,

$$I = \int \phi(x)[1 + \cos 2\pi D(n+x)] \, dx \,.$$

Putting

$$\theta = 2\pi Dn$$
$$P = \int \phi(x) \, dx$$
$$C = \int \phi(x) \cos 2\pi Dx \, dx$$
$$S = \int \phi(x) \sin 2\pi Dx \, dx$$
$$I = P + C \cos \theta - S \sin \theta \,.$$

In order to introduce a quantitative relation between V, the visibility of the interference pattern, and $\phi(x)$, the intensity of the source as a function of the frequency, it is, of course, necessary to assume a definite expression for V.

Let us assume[2]

$$V = \frac{I_1 - I_2}{I_1 + I_2} \,,$$

where I_1 is the intensity at the center of a bright fringe and I_2 that at the middle of an adjacent dark fringe. To find I_1 and I_2, put $dI/d\theta = 0$, giving $\tan \theta = -S/C$, which,[3] substituted in the expression for I, gives

$$I = P \pm \sqrt{C^2 + S^2} \,,$$

[1] In all cases to be considered, the light is so nearly monochromatic that x is small.

[2] The same result follows if we assume $V = \dfrac{dI/d\theta}{I_1 + I_2}$, in which $dI/d\theta$ is the steepest part of the intensity curve.

[3] Since x is always small, C and S may be regarded as constant.

whence

$$V^2 = \frac{C^2 + S^2}{P^2}.$$

Let it be required to find the visibility curve for a few special cases for given values of $\phi(x)$.

Case I.—$\phi(x) =$ constant from $x = -\frac{1}{2}a$ to $x = \frac{1}{2}a$. This being an even symmetrical function, the S integral vanishes and

$$V = \frac{C}{P} = \frac{\sin \pi Da}{\pi Da}.$$

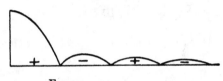

$\phi(x)$ V

FIG. 23

The graph of V is shown in Figure 23. The fringes vanish at equal intervals $D_0 = 1/a$, $2/a$, etc., where

$$a = \partial x = \frac{\partial \lambda}{\lambda^2}.$$

The change of sign of V at alternate loops signifies that the bright and the dark fringes are reversed.

Case II.—$\phi(x) = e^{-x^2/a^2}$. Here also

$$S = 0$$

and

$$V = \frac{\int_{-\infty}^{\infty} e^{-x^2/a^2} \cos 2\pi Dx\, dx}{\int_{-\infty}^{\infty} e^{-x^2/a^2} dx},$$

whence

$$V = e^{-(\pi Da)^2}.$$

Figure 24*b* is of the same form as $\phi(x)$, but its "breadth"[1] is inversely proportional to that of $\phi(x)$.

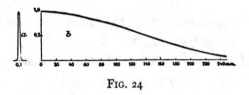

FIG. 24

Case III. Multiple source.—If the spectrum consists of a number of similar symmetrical "lines" whose intensities are proportional to P_n, and whose visibility curve is V, the visibility $\overline{V}$ is given by

$$(\Sigma P)^2 \overline{V}^2 = V^2 \Sigma P_m P_n \cos 2\pi D(x_m - x_n) \ .$$

For a doublet

$$(\Sigma P)^2 \overline{V}^2 = V^2(P_1 + P_2 + 2P_1 P_2 \cos 2\pi Da) \ ,$$

or, if $r = P_1/P_2$, and $L = 1/a$,

$$\overline{V}^2 = V^2 \frac{1 + r^2 + 2r \cos 2\pi D/L}{1 + r^2 + 2r}$$

$$\left(L, \text{ the "period"} = \frac{1}{x_1 - x_2} = \frac{\lambda^2}{\partial\lambda} \right).$$

Figure 22 represents the graph of such a visibility curve for the sodium doublet.

[1] It is convenient to define the "half-width" as the value of the independent variable at which the function falls to one-half its original value. It follows if $x_{1/2}$ is the half-width of the spectral "line," the distance $D_{1/2}$ will be $\dfrac{\log 2}{\pi x_{1/2}}$, or approximately, $D_{1/2} = \dfrac{.22}{x_{1/2}}$.

THE INVERSE PROBLEM

The deduction of the distribution of light in the spectrum of the source is, in general, a problem of considerable difficulty, especially in the case of asymmetrical lines. If this distribution is $I = \phi(x)$, we have, by the Fourier theorem,

$$\phi(x) = \frac{1}{\pi} \int_0^\infty \cos kx \, dk \int_{-\infty}^\infty \phi(u) \cos ku \, du$$

$$+ \frac{1}{\pi} \int_0^\infty \sin kx \, dk \int_{-\infty}^\infty \phi(u) \sin ku \, du,$$

or

$$\phi(x) = \int [C \cos kx \, dk + S \sin kx \, dk],$$

or, since $C = PV \cos \theta$ and $S = PV \sin \theta$,

$$\phi(x) = \int V \cos (kx + \theta) dk.$$

$$(k = 2\pi D)$$
$$V = f_1(k)$$
$$\theta = f_2(k)$$

If ϕ is symmetrical, $\theta = 0$ and $\phi(x) = \int V \cos kx \, dk$. In this case, if V is expressible as an analytical function of k, $\phi(x)$ may be found by integration.[1]

In the case of an asymmetrical function, it is necessary to know both V, the visibility curve, and θ, which may be termed the "phase curve." The latter may be found by comparison with an approximately homogeneous symmetrical source of approximately the same wavelength.

[1] In most of the cases which follow, this integration was effected by means of the harmonic analyzer (see Light Waves, p. 68).

In illustration, consider the two cases

$$\left.\begin{array}{l}\phi(x)=1 \text{ for } x=0\\ \phi(x)=r \text{ for } x=a\end{array}\right\} \quad P_1V_1=\sqrt{1+2r\cos ka+r^2} \quad (1)$$

$$\left.\begin{array}{l}\phi(x)=1 \text{ for } x=0\\ \phi(x)=r/2 \text{ for } x=\pm a\end{array}\right\}PV=1+r\cos ka . \quad (2)$$

For $r=.4$, P_1V_1 is scarcely distinguishable from PV. But, in the first case,

$$\tan\theta=\frac{r\sin ka}{1+r\cos ka},$$

while in the second,

$$\tan\theta=0 .$$

The maximum difference is about $.12\pi$. This corresponds to a shift in the position of the fringes of $\pm.12$ of the distance between the fringes.

For many purposes an exact estimate of the visibility is not essential. For instance, in the investigation of a doublet the distance between the components may be found from

$$\frac{\partial\lambda}{\lambda}=\frac{\lambda}{P},$$

where P is the distance between successive minima.

The ratio of the components may be obtained from

$$r=\frac{V_1-V_2}{V_1+V_2},$$

in which V_1 is the mean of two maxima and V_2 the value of the intervening minimum. This requires an accurate knowledge of V_1 and V_2.

In general, it is necessary to provide comparison fringes whose visibility is defined by

$$V = \frac{I_1 - I_2}{I_1 + I_2}$$

which may be accurately determined. For this purpose a concave crystal of quartz is placed between crossed Nicols, giving circular interference fringes whose intensity is

$$I = 1 - \sin^2 2a \sin^2 \pi\partial ,$$

whence

$$V = \frac{1 - \cos^2 2a}{1 + \cos^2 2a} ,$$

in which a is the angle through which the quartz is rotated to give fringes of distinctness or visibility equal to those shown in the interferometer. (The eye readily learns to dispense with these comparison fringes.)

A few examples follow of some of the results actually obtained in practice.

1. *Hydrogen* $(\lambda = 6563)$.—The visibility curve is represented by the expression

$$V = 2^{-\frac{D^2}{\Delta^2}} \sqrt{\frac{1 + r^2 + 2r \cos 2\pi D/P}{1 + r^2 + 2r}} .$$

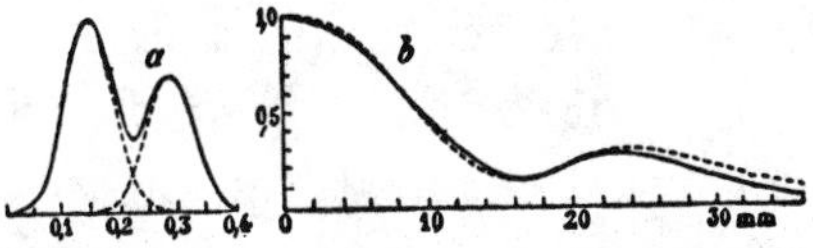

Fig. 25

The line is therefore a doublet (Fig. 25) with components whose half-width is $\partial_1 = \frac{.22}{\Delta}\lambda^2$, and whose distance apart is

$\partial_2 = \dfrac{\lambda^2}{P}$, the ratio of the components being r. For hydrogen in a vacuum tube under a pressure of about 1 mm of mercury, illuminated by discharge from an induction coil, the following values were obtained:

$$\partial_1 = 0.14 \ A.U.$$
$$\partial_2 = 0.05 \ A.U.$$
$$r = 0.70$$

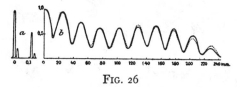

FIG. 26

2. *Thallium* $(\lambda = 5350)$.—The line is a double with components one-tenth A.U. apart, the ratio being 0.5. The visibility curve (Fig. 26) shows that each component is a very close doublet whose distance is only 0.01 A.U.

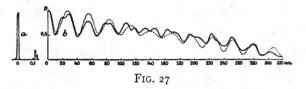

FIG. 27

(This may, however, indicate a reversal, and does not necessarily prove an actual doublet).

3. *Mercury.*—Figure 27 represents the visibility curve for λ 5790, Figure 28 for λ 5770, and Figure 29 for λ 5461. The corresponding distribution of light in the spectrum of the source is given in the figures to the left of the visibility curves, and, while giving results which have had to be somewhat modified by subsequent investigation (chiefly on account of the absence of information of the phase curve), are in the main correct as regards the

intensity and distance of the components (see paper of Gale and Lemon, *Astrophysical Journal*, **31**, 78, 1910.

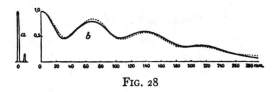

FIG. 28

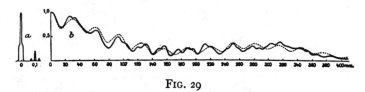

FIG. 29

The following examples illustrate results which show the effect of temperature and pressure on the width of spectral lines.

For low and moderate pressures the principal sources of broadening will be, first, the effect of motion of the vibrating electrons in the line of sight; and, second, the effect of collisions in limiting the free path. If the density is very low, the second effect may be ignored (for example, in the case of hydrogen when the pressure is 1 or 2 mm).

The expression for the visibility as deduced by the late Lord Rayleigh is

$$Vis = exp\left[-\pi\left(\frac{\pi Dv}{\lambda V}\right)^2\right],$$

in which D is the difference in path, λ the mean wavelength, v the square root of the mean square velocity, and V the velocity of light. If Δ be the difference of path corresponding to $vis = \frac{1}{2}$,

$$\Delta/\lambda \doteqdot .15 V/v.$$

Taking $v = 2,000$ m per second, for hydrogen, $\Delta/\lambda = 22,500$.

TABLE I

Substance	Atomic Weight	λ	Δ	$N = \Delta/\lambda$	N (Calculated)
H_r	1	6563	19.0	30,000	22,500
H_b	1	4861	8.5	18,000	22,500
O	16	6160	34.0	55,000	80,000
Na_r	23	6161	66.0	107,000	108,000
Na_y	23	5893	80.0	133,000	108,000
Na_{gy}	23	5676	62.0	109,000	108,000
Na'_g	23	5153	44.0	85,000	108,000
Na''_g	23	4979	55.0	110,000	108,000
Zn_r	65.7	6362	66.0	104,000	182,000
Zn_b	65.7	4810	47.0	98,000	182,000
Cd_r	112.4	6438	138.0	215,000	238,000
Cd_g	112.4	5085	120.0	236,000	238,000
Cd_b	112.4	4800	64.0	134,000	238,000
Hg'_y	200.6	5790	230.0	400,000	317,000
Hg''_y	200.6	5770	154.0	270,000	317,000
Hg_g	200.6	5461	230.0	420,000	317,000
Hg_b	200.6	4358	100.0	230,000	317,000
Tl	204	5350	220.0	400,000	322,000

Again, if we ignore the difference in temperature (about which there is considerable uncertainty) at which other substances were examined, the velocities v should vary inversely as the square root of the atomic weight; and the number of waves in the path difference corresponding to $V = \frac{1}{2}$ is therefore $22,500\sqrt{m}$. Considering the difficulties and uncertainties of the problem, Table I shows a fair agreement between the observed and the calculated values.

A similar investigation at pressures varying from 0 to 100 mm gives as the result (which applies with a fair degree of accuracy to substances varying enormously in density and volatility) the expression

$$\partial = c \sqrt{\frac{\theta}{m}} \, (a+bp) \, ,$$

in which ∂ is the width of the spectral line, θ the absolute temperature, m the atomic weight, p the pressure, $a, b, c,$ constants.

While it may be admitted that the analysis of spectral lines by the method of visibility curves is somewhat indirect and not entirely certain, it has nevertheless proved of considerable value, especially in cases where the effects to be observed are beyond the power of the spectroscope. At the time of its inception the resolving power of the instruments available was far too small for many of the problems which have yielded to the new method, such as the resolution of fine structure, the effect of temperature, of pressure, and of the magnetic field.

While other methods (some of which will be described in a future chapter) have for many purposes superseded the method of light-wave analysis, there are still applications for which the latter process is the most powerful means at our disposal.

CHAPTER V

MEASUREMENT OF STANDARD METER
IN LIGHT-WAVES

As indicated in chapter iv, the measurement of the wave-length of light from an approximately homogeneous source may be made by counting the number of interference fringes which pass during the displacement of one of the mirrors of the interferometer. The possible accuracy of such a measurement is determined by the degree of homogeneity of the source. As was shown in the above-mentioned chapter, this is measured by the difference in path at which interference fringes are still visible, which in some cases was shown to amount to 500,000 light-waves or more. In such a case, the order of accuracy of a measurement may be expected to be something of the order of one part in several million.

It is, of course, essential that the radiation from the source be "simple" or at least separable into simple spectral lines, a condition which was shown to be of somewhat rare occurrence. Among the hundreds of radiations examined none answered the requirement so well as the red radiation from cadmium vapor. The visibility curve for this radiation is expressible with a high degree of accuracy by the formula

$$V = e^{-D^2/\Delta^2},$$

in which D is the difference in path and Δ that corresponding to visibility $1/e$. The graph is given in Figure

24. The constant Δ depends on the temperature and pressure of the vapor and is greatest when these are small. The figure gives $\Delta = 160$ mm, and the interference fringes would still be measurable at $D = 220$ mm. But this distance contains about 350,000 light-waves, or say 700,000 fringes. The optical error of measurement will depend somewhat on the visibility of the fringes but an estimate of one-tenth of the fringe width is quite conservative, and indicates the possibility of making such a measurement to an order of accuracy of one in ten million.

The counting of several hundred thousand fringes is altogether too troublesome and uncertain to be relied upon. Doubtless a number of methods may be suggested of making such a process automatic, and in fact some rather promising attempts in this direction have been made; but the possibility of skipping over one or more fringes through some accident is so serious that a somewhat roundabout and tedious but much surer process was substituted.

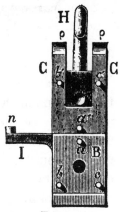

Fig. 30a

This consisted in dividing the distance to be measured (100 mm) into smaller parts so that the number to be counted is much smaller. Thus, if a second intermediate standard of 50 mm is constructed, another of 25 mm, and so on down to one of $100 \div 2^8 = 0.390625$ mm, this last will contain only 600 red light-waves, or 1,200 fringes in the (doubled) distance between the two parallel surfaces constituting the intermediate standard. This number of fringes can readily be counted with certainty. Such an intermediate standard is shown in Figs. 30a and 30b. This consists of two plane-parallel glasses A, A (Fig. 30b), silvered on the front surfaces and held in contact with three brass pins a, b, c (Fig. 30a), which are filed and polished until the two surfaces are as nearly parallel as required.[1] The parallelism is tested in the interferometer arranged as in Figure 31. If mm' represent the front and the rear mirrors of the standard, and d, which may be termed the

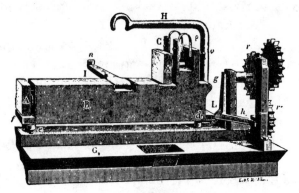

FIG. 30b

"reference plane," is adjusted so as to produce circular fringes in monochromatic light, and if there is no change in diameter of the fringes for m and for m' as the eye is moved across the line of sight, then these are parallel.

The count of the fringes on the smallest of the intermediate standards, which may be designated by I, is made as follows. Let this standard be mm' of the figure, and n an auxiliary mirror adjusted to show circular fringes in red cadmium light. The front surface m is made to coincide with the image of the reference plane, making

[1] The parallelism may be made practically perfect, but an error of a tenth of a micron or so is not objectionable.

with this, however, a small horizontal angle, and giving vertical interference bands in white light, the central band being achromatic and thus readily distinguishable. The reference plane is now moved while at the same time the succession of circular fringes on n are counted. The motion is continued until d coincides with the rear surface m', when vertical fringes appear in white light; the

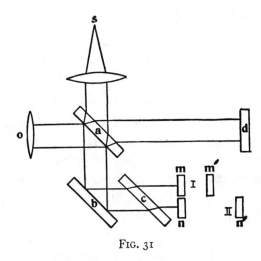

Fig. 31

achromatic band being brought to the same position on m' as it had before on m.

The number of circular fringes counted during the motion will be the whole number required. The fractional excess is very accurately determined by readjusting the reference plane so as to give circular fringes in cadmium light at m and m'. If the phase of the circular fringes is the same at both m and m' then the excess is zero; if this is not the case, the compensating plane-parallel glass c is turned through a very small measured angle until the phase is the same, and by previous calibration the phase difference corresponding to the fractional excess is determined. Such a count made for the three cadmium radiations red, green, and blue[1] gave the numbers shown in Table II.

TABLE II

Red............... 1212.37
Green.............. 1534.79
Blue............... 1626.18

Having the number of waves and the fractional excess in I, the next step is the comparison of I and II. This is effected as follows.

The two standards mm' and nn' are placed side by side and adjusted to parallelism, with the front surfaces nearly in the same plane, so that the slightly inclined reference plane, when properly adjusted, shows vertical fringes on both, the central fringe having the same relative position (say, at the middle) on each face. d is now moved to coincide with m', as shown by the appearance of the vertical fringes on this surface. I is then moved until the fringes reappear on m (and consequently m' and n' are in the same plane or nearly so); and, finally, d is again moved until fringes appear simultaneously on m' and n', the difference being estimated in fractions of a fringe. If this fraction is ϵ, then $II = 2I + \epsilon$. In the present instance, $2 \times 1212.37 = 2424.74$. The fractions are, however, uncertain, and must be corrected by observations with cadmium light as before; the corrected value was found to be 0.93. The same process is repeated in the comparisons of II and III, and so on until the last standard IX

[1] The green and the blue radiations are not quite so simple as the red, but are nevertheless valuable in checking the results.

is reached.[1] In comparing the standards it is not important to know the temperature, provided this is the same for both. But in the measurement of *IX* it is extremely important to know the temperature to within one one-hundredth of a degree, as measured by thermometers carefully tested and whose errors are accurately known.

Table III gives the results of three independent measurements of the number of light-waves in the (doubled) length of standard *IX* (the decimeter). The fact that

TABLE III

Series	Red	Green	Blue
I	310678.48	393307.92	416735.86
II	310678.65	393308.10	416736.07
III	310678.68	393308.09	416736.02

these measurements were made at different times, months apart, and by different individuals and still give the same result to a few hundredths of a light-wave gives confidence in the accuracy of the results.

The final operation is the comparison of the decimeter standard *IX* with the standard meter. For this purpose an auxiliary meter *X* was provided with two diamond scratches at a distance apart very nearly equal to a meter. An arm extending at right angles from *IX* has a similar mark which is placed as nearly as possible in coincidence with one of the meter marks. *IX* is then

[1] Usually the difference between the first fraction and the corrected value is of the order of a tenth of a fringe. There might occasionally be an error as large as 0.2 in which case the doubled value is too near one-half to make it absolutely sure that the doubled whole number is correct. A very valuable check is found in the simultaneous measurements of the green and blue radiations. If the whole number is not right, these show numbers which are quite discordant.

"stepped off" ten times by means of the interferometer fringes. The resulting error is, however, multiplied by ten (instead of by two as in the comparisons of the smaller standards). It is estimated that the error of separate determinations may be of the order of one-half a light-wave. (The mean of all measurements is doubtless much less.) To this error must, however, be added the errors of the micrometric measurements of the "coincidences" at both ends of the meter bar, and finally the errors of the comparison of the auxiliary meter with the standard.[1]

The final result of the determination is given in Table IV.

TABLE IV

Number of Light-Waves of the Three
Principal Cadmium Radiations in
the Standard Meter*

Red.................... 1553163.5
Green.................. 1966249.7
Blue................... 2083372.1

* In air at 15° C. and 760 mm.

It is estimated that these results are correct to the order of one part in two million.[2]

The proposition to make use of the yellow light-wave of sodium as a fundamental standard of length was made many years ago. The method proposed was to measure the angle of diffraction of one of the two sodium radiations produced by a grating. The grating space σ was to

---

[1] A copy of the standard meter. The actual standard is not handled, and is inspected only once in every ten years.

[2] (The relative accuracy for the three radiations is much greater and may correspond to an error of the order of only one part in twenty million.) A similar investigation by Fabry and Pérot, utilizing their well-known "distance pieces," gave results of almost exactly the same value.

be found by a comparison of the end lines of the grating with the standard meter. The relation is given by $\lambda = (\sigma/m) \sin \theta$, where θ is the angle of diffraction, m the order of the spectrum, and λ the length of the light-wave.

The principal difficulties in the way of realizing such a measurement with sufficient accuracy are the following: First, the sodium lines are relatively broad and variable. Second, the measurement of an angle is always more difficult and the results less certain than the measurement of distance. Third, the assumption $\sigma = $ constant on which the formula depends is not always justified. Fourth, the comparison of the distance between the end lines of the grating with the standard meter cannot be carried out with sufficient accuracy. (The attempt was, however, actually made but the result obtained is now known to be in error by about three parts in a hundred thousand.)

Long before this proposal, the need for a so-called "absolute" standard of length was felt, and two standards were proposed which it was hoped would fulfil the requirements. One of these was the length of a pendulum which swings once in a second at Paris. Such a Kater pendulum, consisting of two knife edges with a heavy mass between, is adjusted so that the times of swing are exactly the same when swung from either knife edge, when the length of the equivalent simple pendulum is equal to the measured distance between the knife edges. It was found on trial that the errors of measurement were considerably greater than expected.

A second attempt consisted in the establishment of the forty-millionth part of the earth's circumference as the standard meter, and, in fact, this was legally enacted and furnished the original standard meter. It proved, however, as the results of several of the very costly investigations of the measurement of a given arc of the meridian (from which, and knowing the latitudes of the two ends, the whole circumference could be found) that this measurement was also too inaccurate to serve, so that the actual legal standard today is the quite arbitrary distance between two lines ruled on a bar of an alloy of platinum and iridium. (This alloy possesses hardness and durability in very high degree.)

But though every care may be taken to insure the safety of this arbitrary standard, it is certainly not safe to rely on its permanence when considering quantities to the order of one-millionth. The length of the light-wave of radiations of vapor of cadmium under proper conditions has been proved constant and reproducible at will with an error less than one part in two million.[1]

[1] There is no doubt that the order of accuracy could be still further increased—say, to the order of one in ten million—if it should be thought desirable to make the light-wave (of cadmium vapor, or possibly of some other radiation still more nearly homogeneous) the legal standard.

It would hardly be worth while to attempt this if the purpose be merely to control the length of the present standard which is determined by the distance between relatively coarse and irregular lines, such as would hardly admit of measurements of this high order of accuracy.

It may be mentioned that decimeter standards are now furnished by the Bureau of Standards containing a given number of light-waves.

CHAPTER VI

DIFFRACTION

When a beam of light is limited by an opaque screen with an aperture, the light continues not only in the direction limited by the projection of the borders of the aperture, but also into the geometrical shadow. The phenomenon called "diffraction" is a special case of interference, and is readily explained in its main features by means of Huyghens' principle together with the principle of interference. Thus the effect of any wave-front at a point farther along in the path will be found by summing up the effects of the elementary wavelets whose centers are continuously distributed over the surface of the wave-front. In the general case, this leads to integrals which cannot be evaluated except by processes of approximation.

In the most important case which arises in practice, namely, the investigation of the diffraction figure at the focus of a lens or mirror, the problem is much simplified. Before proceeding to the investigation, however, it is necessary to deduce the expression for the effect of an elementary wavelet. This can be accomplished in elementary fashion as follows.[1] Assume that the amplitude of the effect at the focus is proportional (1) to the area of the element of wave-front dS, (2) inversely proportional to the distance ρ, and (3) independent of the in-

[1] Lord Rayleigh, *Scientific Papers*, Vol. III.

clination. If the disturbance at the original wave-front be

$$V = \cos kat ,$$

where $k = 2\pi/\lambda$, then that at the given point (distant ρ from dS) will be

$$dW = \frac{dS}{\rho} \cos k(at-\rho) ,$$

and the total effect of the entire wave will be

$$\int \int \frac{dS}{\rho} \cos k(at-\rho) ,$$

where the integration is extended over the entire aperture.

To test this result, consider a plane wave of unlimited extent. Let $dS = 2\pi r dr = 2\pi\rho d\rho$, whence

$$W = 2\pi \int_{\rho=f}^{\rho=\infty} \cos k(at-\rho)d\rho ,$$

or

$$W = \lambda \sin k(at-f)$$

instead of

$$W = \cos k(at-f) .$$

In order to obtain the correct result, the disturbance produced by the elementary wavelet should be

$$dW = \frac{-dS}{\lambda\rho} \sin k(at-\rho) . \tag{1}$$

If x, y, z are the co-ordinates of dS and ξ, η, ζ those of the point in the diffraction pattern, then to small quantities of the second order in the focal plane, or to quantities of the fourth order in the focal sphere of radius f,

$$\rho = f - \frac{x\xi + y\eta}{f} . \tag{2}$$

Replacing ds/ρ by $dxdy/f$, and putting $k\xi/f = u$ and $k\eta/f = v$, and counting time from the instant the wave reaches the focus,

$$dW = \frac{-1}{\lambda f} \sin (kat + ux + vy) dxdy ,$$

whence, if $ka = n$

$$\left. \begin{aligned} W &= \frac{1}{\lambda f} \int \int \cos (ux+vy) \cdot \sin nt \, dxdy \\ &+ \frac{1}{\lambda f} \int \int \sin (ux+vy) \cdot \cos nt \, dxdy . \end{aligned} \right\} \tag{3}$$

The preceding formula applies only when the incident wave is constant in amplitude and phase over the entire surface. In the most general case, if Φ represents the amplitude and ψ the phase, both being functions of x and y, then the initial vibration will be represented by

$$V = \Phi \cos (nt + \psi) ,$$

and the corresponding vibration at the focus will be

$$W = -\frac{1}{\lambda f} \int \int \Phi \sin (nt + ux + vy + \psi) dxdy . \tag{4}$$

Putting

$$C = \frac{1}{\lambda f} \int \int \Phi \cos (ux+vy+\psi) \, dxdy$$

$$S = \frac{1}{\lambda f} \int \int \Phi \sin (ux+vy+\psi) \, dxdy \, , \qquad (5)$$

the vibration at the focal plane will be[1]

$$W = S \cos nt - C \sin nt \, , \qquad (6)$$

and the intensity of the diffraction pattern will be

$$I = C^2 + S^2 \, . \qquad (7)$$

Take, as an illustration, $V = \cos nt$ for $x = \pm a$ and zero for all other values.

$$D_x(V) = 2 \cos nt \cos ua$$

$$D_u D_x(V) = \int 2 \cos nt \cos ua \cos uxdu$$

$$= \left[\frac{\sin (a+x)u}{a+x} + \frac{\sin (a-x)u}{a-x} \right] \cos nt \, .$$

If the integration is from $-\infty$ to ∞, the parenthesis is zero except for $x = \pm a$, and for these values it is

$$DD(V) = \cos nt = V \, .$$

[1] If a spherical mirror be made to coincide with the "focal sphere," an image of the source will be formed at the original wave surface. This image may also be considered as the resultant of the disturbance at the focal sphere. Hence if $D(V)$ represents the operation by which W was obtained from V,

$$W = D(V),$$

but by the image

$$V = D(W)$$
$$\therefore \quad V = D[D(V)],$$

an operation which is the exact equivalent of the Fourier formula but in two dimensions (see *Philosophical Magazine* [April, 1905]).

Let it be required to find the diffraction screen which shall produce a given diffraction figure. Let the vibration at the screen be

$$W = \sin nt$$

between $-a$ and a, and zero at all other values. The vibration at the screen will be

$$V = \int_{-a}^{a} [\cos uxdx] \cos nt \cdot dx$$

or for $V_0 = 1$

$$V = \frac{\sin ua}{ua} \cos nt \ .$$

This has negative as well as positive values of the amplitude, and could not be realized by a screen acting by opacity alone.

An appropriate phase factor may be introduced by placing in front of the opacity screen, such as is represented in Figure 32A, a plane-parallel glass plate such as that represented in Figure 32B, in which the alternate rectangles have been etched in hydrofluoric acid so as to introduce a phase difference of half a light-wave.

Figure 33 shows at A the screen which produces the diffraction figure, the graph of which is represented in Figure 32A and an actual photograph of which is shown in Figure 33B. The appearance produced by the combination represented in Figure 32A and B is equivalent to the photograph represented in Figure 33C. With such a combination for a diffracting screen the diffraction figure 33D was obtained.

The formulas giving the relation of the diffraction figure to the diffraction screen are the same in form as those expressing the distribution of light in the spectrum of a non-homogeneous source in terms of the "visibility."

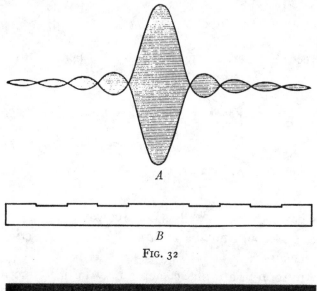

FIG. 32

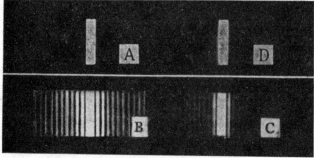

FIG. 33.—*A*, rectangular aperture producing the diffraction figure *B*. *C*, photograph produced by moving *A*, *B* (Fig. 32) vertically. This combination used as a diffracting screen gives at the focus the figure *D*, which is nearly identical with *A*.

Accordingly, if a grating reflects light whose amplitude is the same function of *x* as the visibility curve is of *D*, then the spectrum produced by the grating in homogeneous light will be of the same form as the true spectrum of the non-homogeneous source.

Thus if visibility $= \dfrac{\sin ka}{ka}$, then $I_1 =$ intensity of true spectrum $=$ constant from a to $-a$ and zero for all other values. And if $\phi(x)$ for the grating $= \dfrac{\sin ka}{ka}$ (which may be realized as previously indicated in Figure 32 with an appropriate phase plate), then $I =$ constant from θ to $-\theta$ and zero for all other values. Complete similarity is lacking, however, as there is no phase relation between the elements of the true spectrum.

For Φ and ψ constant ($\Phi = 1$, $\psi = 0$),

$$\lambda fC = \int \int \cos (ux + vy) dx dy$$

and

$$\lambda fS = \int \int \sin (ux + vy) dx dy$$

For an even symmetrical aperture,

and
$$\left. \begin{aligned} \lambda fC &= \int \int \cos ux \, \cos vy dx dy \\ \lambda fS &= 0 \, . \end{aligned} \right\}$$

RECTANGULAR APERTURE

Let $x_1 = a$, $y_1 = b$, giving

$$I = \frac{C^2}{\lambda^2 f^2} = \left[\frac{ab}{\lambda f} \cdot \frac{\sin \frac{1}{2} ua}{\frac{1}{2} ua} \cdot \frac{\sin \frac{1}{2} vb}{\frac{1}{2} vb} \right]^2 .$$

Restoring the values of u and v, and (for small angles) replacing the sine by the arc, putting $a_0 = \lambda/a$, and $\beta_0 = \lambda/b$,

(8)
$$I = \left(\frac{ab}{\lambda f} \frac{\sin \pi \dfrac{a}{a_0}}{\pi \dfrac{a}{a_0}} \frac{\sin \pi \dfrac{\beta}{\beta_0}}{\pi \dfrac{\beta}{\beta_0}} \right)^2 .$$

The intensity is zero for $a = ma_0 = m\lambda/a$ and for $\beta = n\beta_0 = n\lambda/b$. The maxima occur for $\pi\dfrac{a}{a_0} = \tan\pi\dfrac{a}{a_0}$ and for $\pi\dfrac{\beta}{\beta_0} = \tan\pi\dfrac{\beta}{\beta_0}$, giving

for

$$a_1 = 1.43\,\lambda/a \qquad I_1 = 1$$

for

$$a_2 = 2.46\,\lambda/a \qquad I_2 = \left(\frac{2}{3\pi}\right)^2.$$

etc.   etc.

For large m,

$$c_m = (m+\tfrac{1}{2})\pi\lambda/a \qquad I_m = \left[\frac{2}{(2m+1)\pi}\right]^2,$$

with similar values for β_n.

RESOLVING POWER

In the preceding problem, the wave-front incident upon the lens proceeds from a single luminous point and the center of the diffraction pattern corresponds to the geometric image of the point source. If there are two such point sources, as in the case of a double star, each forms its corresponding diffraction pattern, and so long as these are separated by angles large compared with $a_0 = \lambda/a$, there is no difficulty in distinguishing them as a doublet. Lord Rayleigh gives this value[1] as the "limit of resolution."

To find the limit of separation for the microscope, let $2a$ be the angular aperture of the objective at the

[1] If the aperture be circular, as is usual with telescopes, this limit is 1.22 λ/a.

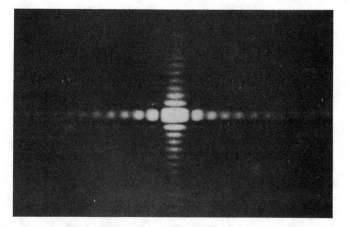

FIG. 34.—Photograph of the diffraction pattern produced by a rectangular aperture.

FIG. 38.—Photograph of the diffraction pattern circular aperture

object and 2β the angle subtended by the objective at the image. Let op (Fig. 35) be the separation of two points

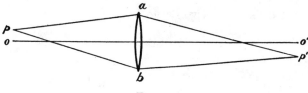

FIG. 35

or lines just resolvable. Then by the Rayleigh criterion the image

$$o'p' = \frac{\lambda}{2 \sin \beta} \cdot$$

But

$$\frac{o'p'}{op} = \frac{\sin \alpha}{\sin \beta} \cdot$$

Hence

$$op = \epsilon = \frac{\lambda}{2 \sin \alpha}$$

for air. If the objective is an immersion lens, the medium having an index of refraction μ, then

$$\epsilon = \frac{\lambda}{2\mu \sin \alpha} \cdot$$

CIRCULAR APERTURE

From the perfect symmetry of the case it is immaterial which radius is taken parallel to u. Accordingly, letting $v = 0$ and integrating for y between limits $\pm \sqrt{R^2 - x^2}$,

$$C = 2R^2 \int_{-1}^{1} \sqrt{1 - \omega^2} \cos n\omega d\omega ,$$

in which $n = \dfrac{2\pi R a}{\lambda} = \pi \dfrac{a}{a_0}$ and $I = \dfrac{C^2}{\lambda^2 f^2}$. The integral C is given in Airy's tables. The values of the maxima and minima are as shown in Table V.

TABLE V

	I	a/a_0		a/a_0	I
First max....	1.000	0	First min...	1.22	0
Second max..	0.01745	1.73	Second min.	2.23	0
Third max...	0.004	2.67			

The graph of I is given in Figure 36, together with that for a rectangular aperture (Fig. 37). Figure 38 (facing p. 62) shows a photograph of the diffraction pattern. The

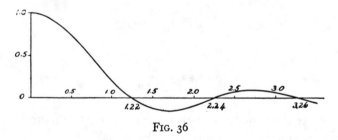

Fig. 36

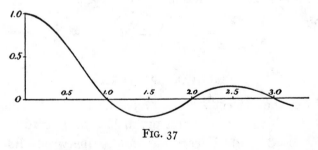

Fig. 37

first minimum occurs at 1.22 λ/a so that the limit of resolution is a little larger with a circular aperture than with a rectangular aperture of the same width.

TWO SIMILAR RECTANGULAR APERTURES

Instead of integrating over the two apertures we may make use of the results already obtained. Since the two apertures are similar, their corresponding points will have a resultant

$$I = 4A^2 \cos^2 \frac{\pi}{\lambda} d \cdot \sin a ,$$

in which $d =$ distance between centers of apertures. The total intensity will be found by giving to A the value already obtained in (8).

For any number n of similar equidistant apertures, $C = \Sigma \cos ux$ and $S = \Sigma \sin ux$, where $x = n\sigma + \beta$, σ being the common distance between the apertures. Converting to exponentials and taking the product $(C+iS)(C-iS)$ and multiplying by A^2, the intensity of the diffraction pattern of a single aperture,

$$I = A^2 \sin^2 n\, \omega / \sin^2 \omega ,$$

where $\omega = \frac{\pi}{\lambda} \sigma \sin \theta$.

Figure 39 shows the graph of the diffraction pattern for $n = 1, 2, 3, 4, 5, 6$, and Figure 40 the corresponding photographs.

The number of cases which might be considered is, of course, infinite. Many of these have been treated both theoretically and experimentally by Fraunhofer, Schwerd, Airy, and others. These results, which were of considerable interest in the earlier days of the discussions, as proofs of the undulatory theory of light, have since lost some of their interest; but some of the experimental

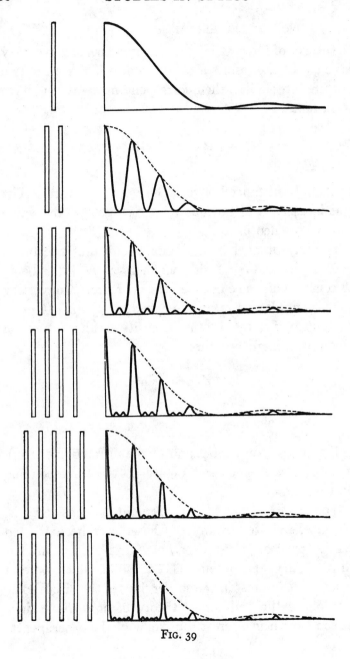

Fig. 39

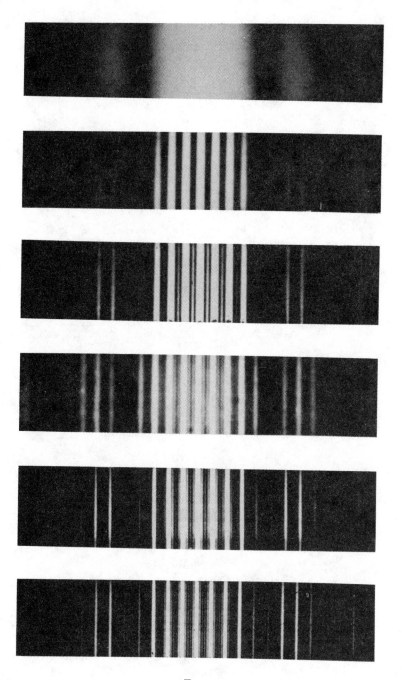

FIG. 40

verifications are in themselves so interesting that a few additional examples are worthy of note.

The general expression for a number of similar point apertures distributed in any manner is

$$\lambda^2 f^2 I = R^2 = C^2 + S^2 \text{ , in which}$$

$C = \Sigma r \cos (ux + vy + \omega)$ and $S = \Sigma r \sin (ux + vy + \omega)$, r and ω being functions of x and y. If phase and intensity are both constant,

$$C = \Sigma \cos (ux + vy)$$
$$S = \Sigma \sin (ux + vy) .$$

For finite apertures similar and similarly oriented, the results are to be multiplied by A^2, the intensity of the diffraction pattern due to a single aperture.

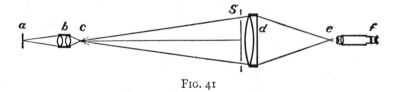

FIG. 41

Following is the simple arrangement actually employed in the observations. Light from a tungsten lamp or an arc is concentrated on a pinhole a, Fig. 41, an image of which is formed by a microscope objective b at c, whence the light proceeds to the objective of a telescope d of 16-inch focal length, forming an image of c at e which is viewed by a low-power microscope at f. The diffracting screen is placed in front of the objective at s.

The accompanying figures are photographs of the diffraction figures D obtained with a number of screens S, together with the analytical expression for the intensity.

If n similar apertures are irregularly distributed and sufficiently numerous, the resulting pattern is the same as that for a single aperture but n times as bright.

BABINET'S THEOREM

Let I_1 be the intensity of any diffraction pattern and I_2 the pattern due to a screen in which the opaque and the transparent parts are interchanged,

$$I_1 = C_1^2 + S_1^2$$
$$I_2 = C_2^2 + S_2^2 .$$

The intensity of the diffraction pattern not too near the center of the field is approximately zero if the screen is removed. Accordingly,

$$I_0 = (C_1 + C_2)^2 + (S_1 + S_2)^2 = 0$$
$$\therefore C_1 = -C_2 \; ; \; S_1 = -S_2 .$$

Hence

$$I_1 = I_2 .$$

The intensity of the diffraction pattern is the same for all points except those very near the center. Thus, if a glass plate be covered with lycopodium dust, the particles being approximately spheres of equal diameter, the resulting diffraction figure would be the same (except at the center) as that produced by a similar distribution of circular apertures of equal diameter. If the diameter of the globules is d, the angular semidiameter of the first dark ring will be $a = 1.22 \; \lambda/d$, from which d may be found from the measured value of a in monochromatic light of wave-length λ.

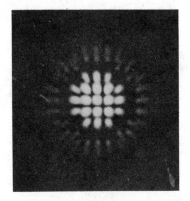 

$$I = (\cos ua \cos vb)^2 A_1^2$$

Fig. 42

 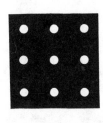

$$I = (1 + 2 \cos ua)^2 (1 + 2 \cos vb)^2 A_1^2$$

Fig. 43

(It is often asserted that the "coronas" occasionally observed when light fleecy clouds pass over the sun or the moon are to be explained in the same way. There may be cases of this kind, but such coronas are much more frequently due to ice crystals. In fact, the succession of colors is the reverse of those due to diffraction.)

EFFECT OF IMPERFECTIONS IN OPTICAL INSTRUMENTS

The preceding applications of formulas (5) assume a uniform intensity and phase over the spherical wave surface. The only case likely to arise in practice in which the intensity is not constant is $\phi(x) = e^{-cx}$, as in the case of the prism spectroscope. Here $C = \int e^{-cx} \cos uxdx$ and $S = \int e^{-cx} \sin uxdx$. If the integral be taken from 0 to ∞, the result,

$$\lambda^2 f^2 I = C^2 + S^2 = \frac{c^2}{c^2 + u^2},$$

is sufficiently near for illustration. Thus, if the intensity of the pencil of rays passing through the base of the prism be 0.018 of that passing through the apex, $ca = 4$,

$$\therefore \frac{I}{I_0} = \frac{1}{1 + \left(\frac{1}{2} \cdot \frac{\pi}{\lambda} a \sin \theta\right)^2}.$$

For $\sin \theta = \lambda/a$ (limit of resolution), $I/I_0 \doteq \frac{2}{7}$ (instead of zero); therefore the resolving power is diminished.

Of considerably greater importance are the imperfections due to a departure of the wave form from perfect sphericity. This is equivalent to a variation in phase which must now be expressed as a given function ψ of x and y. The resulting diffraction figure is found by means of formulas (5), which, however, are simplified by the omission of the amplitude factor. Except in the case just considered, this is usually permissible.

Accordingly, if the vibration at the wave surface be represented by

$$V = \cos(nt - \psi),$$

the intensity of the diffraction pattern will be given by

$$\lambda^2 f^2 I = C^2 + S^2$$

in which

$$C = \int\int \cos(ux + vy - \psi)dxdy$$

$$S = \int\int \sin(ux + vy - \psi)dxdy.$$

If ϵ is the departure from sphericity, $\psi = f(x, y) = 2\pi \epsilon/\lambda$.

The following simple cases may be treated without actually effecting the integration: (1) If $\epsilon = ax$, the result is equivalent to a change in the direction of the wave-front equal to a. Hence the diffraction figure is unaltered save for this displacement. (2) Let $\epsilon = \beta x^2$. For a first approximation let the wave surface be given by $y = x^2/2R$, whence $\delta y = \epsilon = \dfrac{x^2}{2} \delta \dfrac{1}{R}$, and $\delta \dfrac{1}{R} = 2\beta$. The diffraction figure is unaltered if the focus be changed by an amount $\delta R = 2\beta R^2$.

RAYLEIGH'S LIMIT

From the mode of manufacture of optical surfaces these are, in general, surfaces of revolution. Accordingly, only even powers of x need be considered in the expression for y. Omitting powers higher than x^4,

$$y = ax^2 + bx^4.$$

 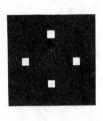

$$I = (\cos ua + \cos vb)^2 A^2$$

Fig. 44

$$I = (1 + 4 \cos^2 ua + 4 \cos ua \cos \sqrt{3}va)A^2$$

Fig. 45

The corresponding equation to the generating circle of the spherical wave surface is

$$y_0 = R - \sqrt{R^2 - x^2} \, ,$$

or, to the same approximation,

$$y = \frac{x^2}{2R} - \frac{x^4}{8R^3} \, .$$

If now $a = 1/2R$, the difference

$$\epsilon = \left(b + \frac{1}{8R^3}\right)x^4 \, ,$$

and the corresponding difference in phase, $\psi = hx^4$.

Table VI gives the corresponding values of I at the

TABLE VI

h	I
0.................	1.00
$\pi/4$.............	$.95$
$\pi/2$.............	$.80$
π...............	$.39$

center of the diffraction figure for given values of h. The loss of light for $h = \pi/2$ is only 20 per cent, which, in most cases, according to Rayleigh, may be tolerated. But $h = 2\pi\epsilon/\lambda = \pi/2$, whence the permissible error, $\epsilon = \lambda/4$.

It will be found in practice that this limit of tolerance is sufficient in most cases in the usual employment of lenses, mirrors, and prisms if the errors are not of a systematic character and in particular if they are not periodic. (For such surfaces as are used in the interferometer this limit is much too large.)

The limit $\lambda/4$ applies to the wave-front. If this is produced by reflection, the corresponding limit for the reflecting surface for normal incidence is $\lambda/8$, and for a refracting surface, $\dfrac{\lambda}{4(\mu-1)}$. Hence the advantage of the refractor over the reflector if there were no other considerations (such as chromatic aberration of refractors and their greater cost).

If the incidence is at an angle i, it can readily be shown that

$$\epsilon = \frac{-\lambda}{4(\cos i - \mu \cos r)},$$

which for normal incidence gives $\dfrac{\lambda}{4(\mu-1)}$ for a refracting surface (as above). For a reflector we have merely to put $\mu = -1$, giving

$$\epsilon = \frac{\lambda}{8 \cos i}.$$

CHAPTER VII

TESTING OF OPTICAL SURFACES

PLANE SURFACES

These are tested by placing on the surface a test plane which is supposed correct to at least the order of accuracy required. The light from a Cooper-Hewitt lamp falls on the surfaces at an incidence approximately normal, and the interference fringes are observed. If these are straight in all azimuths of the test plane relatively to the plane to be investigated, the latter is a true plane to the same order as the test plane. It is possible to detect an error of about 2 per cent of the error corresponding to one fringe; that is, of the order of a hundredth of a light-wave (a few tenths of 1 per cent is possible under favorable conditions and with some practice. In these tests time must be allowed for complete equalization of the temperature; an hour or more may be necessary.)

The manufacture of the test plane requires three working surfaces. These are tested as follows: *A* and *B* are polished to "fit," that is, so that the interference fringes are straight in all positions. This means that the surfaces have equal and opposite curvatures. Next *A* and *C* are made to fit, and also *B* and *C*. This process is repeated until all three fits are exact, in which case all three surfaces must be plane. Figure 46 shows the advancing improvement in a piece of ordinary window glass in the process of correction.

PLANE-PARALLEL PLATES

The surfaces being already very nearly plane parallel, if illuminated by light from a Cooper-Hewitt lamp or

other source of approximately homogeneous radiations, show circular interference bands which may be observed either by the unaided eye, or better, by a low-power telescope.

If the surfaces are not strictly parallel, there will be a change in the diameter of the circles as the plate is moved parallel with its surface. These are accordingly corrected by local retouching until there is no perceptible change.

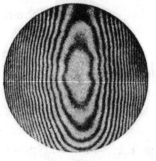

TESTING OF ANGLES

In a recent modification of the method of the measurement of the velocity of light, it was necessary to provide an octagonal revolving mirror in which the angles were to be correct to within one part in a million. This was accomplished by the use of the test angle described on page 134, in chapter xii on the Velocity of Light.

FIG. 46

SPHERICAL SURFACES

The surface to be tested is illuminated by light from a pinhole at *s* (Fig. 47), forming an image at *e* which is just obstructed by the knife edge at *e*. If the eye be

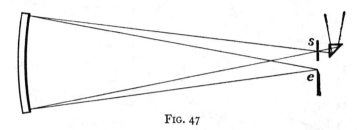

FIG. 47

placed at *e*, all parts of the surface should be dark. But any departure from sphericity will be manifested by corresponding illumination at the defective part.

In the case of a telescope objective, a plane mirror is placed behind the objective, as in Figure 48. The same

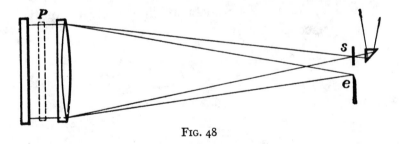

FIG. 48

arrangement is employed for the detection of defects in homogeneity of optical media (striae, etc.), the specimen in the shape of a plane-parallel plate being placed as in *P* of Figure 48.

For astronomical reflectors, the surface must be paraboloidal and the test is modified accordingly, as shown in

Figure 49. The light from the pinhole at *s* falls on the paraboloid at *P*, whence it is reflected to the plane mirror at *R*, retracing its path approximately so as to form the image at the knife edge *e*. There will be a certain astig-

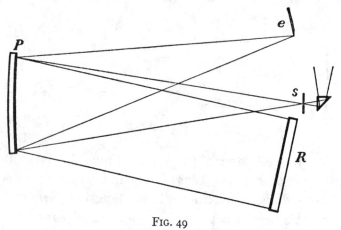

FIG. 49

matism in such an arrangement, which may be obviated by making *R* a plane-parallel plate lightly silvered on the front surface, as in Figure 50.

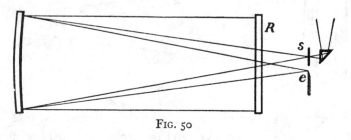

FIG. 50

INTERFERENCE TEST

Light from a Nernst glower is concentrated on a slit *s* (Fig. 51) by a microscope objective[1] *o* whence by a total

---

[1] The microscope objective should be tested and used only if practically free from defects.

reflection prism it proceeds to the spherical surface to be tested. The image of the slit is formed immediately above the prism, and is viewed by a microscope with a 6-mm objective. A series of screens with two aper-

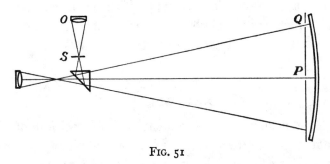

FIG. 51

tures Q and P are placed before the mirror, P over the center, and Q at varying distances from the center.

Interference bands are observed in the microscope, the central band corresponding exactly with the center

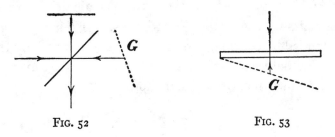

FIG. 52   FIG. 53

of the slit image if the mirror is perfect. The distance between the center of the band and the slit image in fractions of the band-width gives twice the error of the surface at Q.

The same process may be applied to a lens by backing the lens with a plane mirror. With evident modifications, the same method applies to prisms and gratings, in which case the source of light must, however, be approximately monochromatic. In this case, however, the interferometer method as shown in Figure 52, or even more simply, as in Figure 53, may be applied. In either case the interference fringes are concentric circles; and if the grating is perfect, the diameter of the circles remains constant when the eye is moved in any direction over the surface.

ON THE LIMIT OF ACCURACY IN OPTICAL MEASUREMENT

Any accurate measurement of the position of the image of a minute source, such as a pinhole or a narrow slit, or a line ruled on glass or metal (which may be bright on dark ground or the reverse), is effected by adjusting the crosshair of a micrometer to "coincidence" with the image. One of the simplest methods of making coincidence is to bisect the image by the crosshair. Another method is to bisect the distance between a pair of parallel crosshairs by the image. In either case, if β is the angle subtended by the image (or, in the second method, by the distance between the crosshairs), and $\delta\beta$ is the average error of a setting, experiment shows that

$$\partial\beta = b + c\beta , \qquad (1)$$

in which b and c are constants which vary somewhat with the observer, and even with the same observer on different occasions. (In my own case, $b \doteq 5''$ with occasional values as low as $2''$, and $c = 0.0025$.)

It may be noted that the observations are very uncertain, especially if the number is small, the most important difficulty being that of maintaining the constancy of the particular phase of the diffraction image on which the setting is made. The image is often slightly unsymmetrical, which makes the result still more uncertain. Another serious complication arises from the circumstance that the attention and care in making a setting will be apt to vary (for example, by fatigue) during the progress of the series, with resulting errors much larger than the average.

Formula (1) may be applied to either telescope or microscope. In either case, let F be the distance of the image from the objective, f the focal length of the objective, and f_1 that of the eyepiece.

TELESCOPE

If a is the apparent diameter of the image as viewed from the objective, and ∂a the average error of setting,

$$\partial a = \frac{f_1}{F}(b + c\beta) ,$$

or

$$\partial a = \frac{f_1}{F}b + ca .$$

$$(2)$$

(In the case of a point source $a = 2a_0 = 2.44\lambda/a$, where a is the diameter of the objective.)

From this it follows that to obtain the highest attainable accuracy, the magnification $M = F/f_1$ should be considerably greater than $\frac{b}{2ca_0}$, say $\frac{3b}{2ca_0}$. Thus, for the 100 inch at Mount Wilson, $a_0 = 0''.05$, whence $M = 60,000$. Clearly such a high power would be impracticable, both on account of the weakening of illumination and of the very limited field and consequent difficulty of keeping the telescope properly pointed. A magnification of 3,000, however, is not unusual. For this, and with perfect seeing,[1]

$$\partial a = 0.''002 .$$

(The probable error would be about one-fourth as large, or $0.''0005$.)

There are two sources of error somewhat difficult to allow for. The first is that due to faulty focusing which introduces errors of parallax, which may in ordinary work be extremely small, but which cannot be ignored in measurements of highest precision. The second is that due to the motion of the image due to atmospheric disturbances. Such can hardly be attributed to poor "seeing"; yet it is, of course, a modification of the confusion produced by the same cause. In the former case, there is relatively slow motion of the sharp image, while, in the second, there is an integration in time or space or both which produces confusion. Such effects may also be due to vibrations or actual displacements.

In the following example, the objective consisted of a pair of 6-inch objectives of 100-inch focal length, made by G. W. Hewitt and presented by Elihu Thompson. The source was a narrow slit illuminated by a tungsten lamp, the light being filtered through a blue-green gelatin film. The source, placed at 100 inches from the pair of objectives, gave an image at an equal distance beyond, which was viewed by eyepieces giving magnifications of 50, 100,

---

[1] If d is the reduced diameter of the objective necessary to show diffraction rings, then this is the value to be used instead of a. For the utilization of such a diaphragm for measuring the "seeing," see *Yearbook of the Carnegie Institution* (1922), p. 245.

200, and 300. The results for the average error of a setting agreed with the values given by

$$\partial a = \frac{10''}{M} + .005\,a_0 ,$$

to quantities of the order of $0''.01$. M is the magnification and $a_0 = 1.22\lambda/a$.

THE MICROSCOPE

If, in formula (2), we substitute ϵ/f for a, in which $\epsilon = \epsilon_0 + \epsilon_2$, and $\epsilon_0 =$ twice the limit of resolution, or $\dfrac{\lambda}{\sin \phi/2}$, $\epsilon_2 =$ diameter of object, and $f =$ distance from object to border of objective,

$$\partial \epsilon = \frac{f f_1}{F} b + c\epsilon , \tag{3}$$

or

$$\partial \epsilon = \frac{f_1}{m} b + c\epsilon , \quad \left(m = \frac{\mu \sin \phi/2}{\sin \psi/2} \right) ,$$

$\phi =$ angular aperture from object, $\psi =$ angular aperture from image.

This expression for the average error in setting may also be used for an independent determination of the constants. A series of micrometric measurements with varying objectives and eyepieces gave $b = 6''$ and $c = 0.003$, in good agreement with the results obtained with the telescope. Among these was one with $a = 8$ mm, $f = 6.5$ mm, $f_1 = 16$ mm, and $F = 200$ mm, for which, therefore, $\epsilon(= \epsilon_0) = \dfrac{\lambda}{\sin \phi/2} \doteq 1.0$, giving

$$\partial \epsilon_{calc} = 0''.019$$

while

$$\partial \epsilon_{obs} = 0''.020 .$$

THE SIGHT TUBE

Finally, these constants may be determined by the "sight tube" without any of the complications of the lenses. A tube a meter in length was provided at the eye end with a circular diaphragm 1 mm in diameter, and at the other end with a pair of sight marks of the form indicated (Fig. 54). The tube was mounted on a vertical axis and the sights were made to "bisect" a distant illuminated circular aperture, the angle of the sight tube being read off by mirror and telescope. The final result of such a series of measurements gave $b = 2''.8$. The prob-

FIG. 54

able error of naked-eye observations with the sight tube is about one second of arc.

THE INTERFEROMETER

The interferometer as ordinarily employed produces interference bands whose width, $w = \lambda/\phi$, where ϕ is the inclination of the two interfering pencils. Two sets of measurements (the first at the bisection of a dark fringe, and the second at the bisection of a bright fringe) gave the following results:

Dark Fringe	Bright Fringe
$c = \dfrac{\delta w}{w} = .011$	$c = .0057$

In a very bright light, the physiological effect (as also the photographic effect on overexposure) is such as to diminish the apparent width (of the dark fringe in the former case, and the bright fringe in the latter). Under these conditions a second series of measurements gave

$$c = .0035 .$$

A still higher degree of accuracy is obtained by repeated reflections produced by the two half-silvered plates as in the Fabry-Perot apparatus. Three trials were made with such an arrangement, the first with a chemical silver deposit transmitting only 1 per cent of the light. The second was a sputtered gold deposit of about the same transmission, while the third was a sputtered silver deposit transmitting about 10 per cent. These gave as the mean value

$$c = .0025 .$$

The average error in the angular position of the central fringe is

$$\partial\beta = b + c\beta ,$$

where $\beta = rw/f$ ($r = 1$ for a single reflection; but as observed in a very bright light, or in overexposed photographs, r is less than unity; and with repeated reflections, as in the Fabry-Perot apparatus, r may be of the order of .05). The fringe width $w = \lambda/\phi$, where ϕ is the inclination of the two interfering pencils. Hence

$$\beta = \frac{r\lambda}{\phi f_1} .$$

If $\delta\epsilon$ is the average error in distance,

$$\partial\epsilon = \phi f_1 \partial\beta = b f_1 \phi + cr\lambda .$$

But ϕ may be made as small as desired, whence

$$\partial\epsilon = .0025 r\lambda .$$

With only one reflection, and for $\lambda = 0^\mu 6$,

$$\partial\epsilon = 0^\mu 0015 \, ,$$

while with repeated reflections $(r = .05)$, $\delta\epsilon = 0^\mu 000075$.

In the form of apparatus which permits of multiple reflections (in which as many as twenty have been observed), the results of actual measurement gave for the constant c the value .015 in the tenth order. This would correspond to an average error of $\dfrac{0.015}{40} \lambda$, or approximately

$$\partial\epsilon = 0^\mu 0002 \, .$$

With carefully polished surfaces it is estimated that c may be as small as .003, corresponding to an average error

$$\partial\epsilon = 0^\mu 00004 \, ,$$

or one twenty-five millionth of a millimeter.

Repeated reflections may also be utilized to magnify a change in the difference of path between two interfering pencils of light, and thereby to measure very small displacements. An application of this principle is illustrated in Figure 55, which represents the mirror system of a spring gravity meter. The heavy lines indicate the silvered portions of the mirrors, which are so arranged that the beam of light is divided at a, one portion being reflected many times in the upper space, and the other in the lower space. The two beams are reunited at b. The plate P is mounted on a long quartz spring lever. A slight downward motion of P results in an increase of the upper path and a decrease of the lower path, with a resultant displacement of the interference fringes.

The number of reflections may be increased as much as twenty fold, with a corresponding increase in the accuracy of measurement.

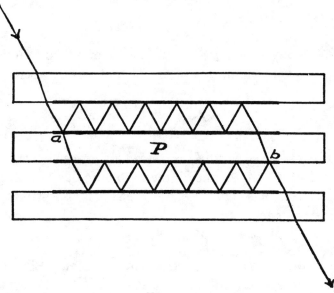

FIG. 55

CHAPTER VIII

DIFFRACTION GRATINGS

The expression for the intensity of the diffraction figure for any number n of equidistant similar apertures is

$$I = A \frac{\sin^2 n\omega}{\sin^2 \omega},$$

in which $\omega = \pi n\sigma/\lambda \sin \theta$, $\sigma = $ distance between apertures, and $A = $ intensity of the diffraction figure due to a single aperture. If n is very large, the intensity is approximately zero except for

$$\sin \theta = m\lambda/\sigma .$$

If the apertures are long and narrow rectangular slits, the screen is called a "diffraction grating."

The incident light is diffracted in directions θ depending on the wave-length, and is thus analyzed into its constituent colors in the form of a spectrum. The incident wave-front is supposed to coincide with the (plane) diffraction screen or grating. If the plane wave-front is inclined at an angle i, the formula becomes

$$\sin i + \sin \theta = m\lambda/\sigma . \tag{1}$$

The incident wave emanates from a point source; but the result is the same, except for intensity, if the source be a very narrow slit parallel with the length of the apertures. In practice, the slit is in the focus of a collimating telescope from which the light issues as a parallel beam and is diffracted as a parallel beam which, by an observing or a photographic objective, is brought to a focus where, in monochromatic light, it forms an image of the slit. The resulting spectrum is thus a succession of such slit images.

The change in the angle θ corresponding to a given change in wave-length is called the dispersion $D = d\theta/d\lambda$. Differentiating equation (1) for $i = $ constant,

$$\frac{d\theta}{d\lambda} = \frac{m}{\sigma \cos \theta},$$

or, if l is the length of the grating and a the width of the beam of light entering the telescope, and if the grating space σ is replaced by the reciprocal of n_1, the number of apertures per centimeter,

$$\frac{d\theta}{d\lambda} = (mn_1)l/a . \tag{2}$$

The resolving power may be defined as the reciprocal of the ratio $\delta\lambda/\lambda$ where $\delta\lambda$ is the smallest difference in wave-length which can be resolved, or

$$R = \frac{\lambda}{\delta\lambda} . \tag{3}$$

The limit of resolution of the observing telescope (of rectilinear aperture a) is $\delta\theta = \lambda/a$, which in equation (2) gives

$$R = (mn_1)l , \tag{3a}$$

or if the total number of apertures is $n = n_1 l$,

$$R = mn . \tag{3b}$$

From equations (2) and (3a),

$$R = aD .\qquad(3c)$$

Finally, if the value of m from (1) is substituted in (3b),

$$R = \frac{l}{\lambda}(\sin i + \sin \theta) .\qquad(3d)$$

The maximum resolving power is attained for grazing incidence and diffraction; but in this case the intensity would be vanishingly small. Practically the maximum resolving power may be taken as

$$\bar{R} = 1.75\, l/\lambda ,\qquad(3e)$$

and is independent of the grating space. Thus, for a grating $l = 200$ mm long and for light of wave-length $\lambda = .0005$ mm,

$$\bar{R} = 700,000 .$$

Such a grating would separate the components of a doublet of only one seven-hundredth of the distance between the sodium lines.

DISPERSION AND RESOLVING POWER OF PRISMS

The difference in optical path of the extreme portions of a wave-front originally and ultimately plane may be represented by[1]

$$\int \partial\mu ds - \int \partial\mu ds_1 = a\partial\theta .$$

If μ, the index of refraction, is constant for the two paths as in the case of prisms,

$$\partial\theta = \partial\mu(s - s_1)/a ,$$

[1] *Rayleigh Scientific Papers.*

or if l is the base of the prism,

$$\partial\theta/\partial\mu = l/a = \frac{d\theta}{d\lambda}\frac{d\lambda}{d\mu} ,$$

whence the dispersion

$$D = \frac{d\theta}{d\lambda} = \frac{l}{a}\frac{d\mu}{d\lambda} . \tag{4}$$

The condition for the limit of resolution is $\delta\theta = \lambda/a$, which gives

$$R = \frac{\lambda}{d\lambda} = l\frac{d\mu}{d\lambda} . \tag{5}$$

Let it be proposed to find the number of prisms of dense flint glass which gives the same resolving power as a grating whose length is the same as that of the base of one prism. Equating the values of R in $(3e)$ and (5) and replacing l by NL in (5) gives $N\lambda\, d\mu/d\lambda = 1.75$. For dense flint glass $\lambda d\mu/d\lambda = 0.11$ whence $N = 16$.

BRIGHTNESS OF GRATING SPECTRA

For a grating acting by opacity only, the intensity is expressed by

$$I = \frac{\sin^2 (\pi \sin \theta\, a/\lambda)}{(\pi \sin \theta\, a/\lambda)^2} ,$$

but $\sin \theta = m\lambda/\sigma$, hence

$$I = \frac{\sin^2 m\pi\, a/\sigma}{(m\pi\, a/\sigma)^2} .$$

For a grating in which the opaque and the reflecting parts are of equal width, $a = \sigma/2$, so that

$$I = \frac{\sin^2 \frac{1}{2}m\pi}{(\frac{1}{2}m\pi)^2} .$$

If the intensity of the central image be taken as unity, the intensities of the successive orders will be approximately 0.4, 0.0, 0.044, 0.0, 0.016, etc., showing the rapid falling off of intensity in the spectra of higher order.

This circumstance would have to be taken into account in the comparison between the performance of the grating and that of the prism. It would be of great advantage to observe in a low order. Let this be the first $(m=1)$. The resolving power R would then be n, the number of ruled lines, whence $n=1.75\ l/\lambda$, which would give for the grating space $\sigma=\lambda/1.75$. Gratings with rulings as close as this would, however, be difficult to manufacture.[1]

But gratings act by "retardation of phase" as well as by "opacity," and in some cases the former is the essential factor, and for such gratings the preceding result may be profoundly modified.

PHASE GRATINGS

General case.—If the vibration at the diffraction screen is $V=\phi\ \sin nt+\psi\ \cos nt$, the resulting disturbance at the focus will be

$$W=\int \phi\ \sin (nt-ux)\ dx+\int \psi\ \cos (nt-ux)dx$$

or

$$W=\sin nt\left[\int \phi\ \cos ux\ dx+\int \psi\ \sin ux\ dx\right]$$
$$-\cos nt\left[\int \phi\ \sin ux\ dx-\int \psi\ \cos ux\ dx\right].$$

The intensity of the diffraction pattern is therefore

$$I=\left[\int \phi\ \cos ux\ dx+\int \psi\ \sin ux\ dx\right]^2$$
$$+\left[\int \phi\ \sin ux\ dx-\int \psi\ \cos ux\ dx\right]^2.$$

[1] A very good 25-mm grating has been ruled with 2,000 lines per mm.

If the diffraction screen is a grating of period $\sigma = 2\pi/\kappa$, ϕ and ψ may be represented in Fourier series as

$$\phi = \Sigma a_m \sin m\kappa x$$
$$\psi = \Sigma b_m \cos m\kappa x .$$

The resulting spectrum of mth order will have the intensity given by

$$\sqrt{I_m} = a_m \int \cos (m\kappa + u)x \, dx + a_m \int \cos (m\kappa - u)x \, dx$$
$$+ b_m \int \cos (m\kappa + u)x \, dx + b_m \int \cos (m\kappa - u)x \, dx .$$

The integrals vanish except for

$u = m\kappa$ for which $\qquad I_m = (a_m - b_m)^2$,

and

$u = -m\kappa$ for which $\qquad I_m = (a_m + b_m)^2$.

For the former, $\qquad \sin \theta = m \lambda/\sigma$;

for the latter, $\qquad \sin \theta = -m \lambda/\sigma$.

If now all the spectra are to vanish except the mth, then

$$\tan \omega = \tan m\kappa x$$

$$\omega = m\kappa x + 2n\pi = \frac{2\pi}{\lambda} 2y \cos^2 \theta/2 ,$$

where y is the depression in the surface produced by the ruling tool. But from $\sin \theta = m\lambda/\sigma$, this becomes

$$\sin \theta \left(x + \frac{n}{m} \sigma \right) = 2y \cos^2 \theta/2 ,$$

whence

$$y = \left(x + \frac{n}{m} \sigma \right) \tan \theta/2 ,$$

the graph of which is shown in Figure 56. (This result might have been anticipated from the circumstance that

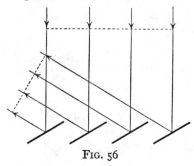

in order to reflect the normally incident light at the angle θ the reflecting surface must be inclined to the plane of the grating at the angle $\theta/2$.)

FIG. 56

Attempts to realize gratings of this character have generally met with only moderate success. For this purpose, a fragment of diamond is selected of appropriate shape and properly inclined. The resulting rulings frequently give spectra which are abnormally bright in one order. In consequence of the wearing away of the cutting edge, however, large gratings of this character are rather rare. It will be shown in the next chapter how the required result may be obtained by the "echelon."

EFFECT OF ERRORS OF DIFFRACTION GRATINGS

The permissible error in ruling may be found from the Rayleigh limit. If this be represented by ϵ, then

$$\epsilon = \frac{\lambda}{4(\sin i + \sin \theta)} = \frac{\sigma}{4m}.$$

In practice it is convenient to distinguish three types of error. These, together with an approximate estimation of the corresponding permissible error in ruling are

Irregular.................1.00 λ
Systematic............... .20 λ
Periodic .... .05 λ

Such estimates will of course depend on the character of the systematic error, and a more precise value will be found by an investigation of the character of the resulting spectrum.

The most general expressions for the amplitude of the light at the diffracting surface is

$$\Phi = f(x) \cos(\kappa x - \beta) ,$$

where $\kappa = \dfrac{2\pi}{\sigma}$, and $f(x)$ and β are both functions of x. If this be substituted in the general formula for the diffraction pattern

$$C = \int f(x) \cos(\kappa x - \beta) \cos(ux - \psi) \, dx ,$$

or

$$C = \int f(x) \cos[(\kappa - u)x - (\beta - \psi)] \, dx$$

and

$$S = \int f(x) \sin[(\kappa - u)x - (\beta - \psi)] \, dx$$

(omitting the spectrum corresponding to $[\kappa + u]$).

But these formulae are the same as those of (5) (p. 58) (Diffraction) if in these we put

$$f(x) \text{ for } \Phi$$
$$\beta - \psi \text{ for } \psi$$
$$u - \kappa \text{ for } u$$
$$v = 0 .$$

But

$$u - \kappa = \frac{2\pi}{\lambda} \sin\theta - \frac{2\pi}{\sigma} = \frac{2\pi}{\lambda}(\sin\theta - \sin\theta_0)$$

$$= \frac{2\pi}{\lambda} \cos\theta \partial\theta.$$

This is accordingly the value to be substituted for u in the expressions

$$C = \int f(x) \cos ux \, dx$$

$$S = \int f(x) \sin ux \, dx .$$

As examples of the application of these expressions, consider, first, the grating $\phi = e^{-cx} \cos \kappa x$.

$$C = \int_0^l e^{-cx} \cos ux \, dx$$

$$S = \int_0^l e^{-cx} \sin ux \, dx$$

from which the approximate value of the intensity of the diffraction image is

$$\frac{c^2}{(c^2 + u^2)^2}$$

in which

$$ul = \frac{2\pi}{\lambda} l \cos \theta \partial\theta \; ;$$

or since $l \cos \theta = a$, the width of the beam entering the observing telescope, and putting $a_0 = \lambda/a$ and $a = \delta\theta =$ angular distance from $\theta = m\lambda/\sigma$,

$$ul = 2\pi a/a_0 .$$

Figure 57 gives the graph of I for $c = 4/l$ (dotted line) together with that for $c = 0$ (full line), and shows that the resolving power in the former case is roughly three-fourths of that of the latter.

Take, as a second example, a kind of error which is apt to occur in consequence of the accumulation of pressure of the driving mechanism in ruling engines and its sudden release, thus producing a sudden change of phase without change of spacing in the remaining parts of the grating. There may be several such changes, in which case the resulting grating would be no better than one whose length is the average of the lengths of constant spacing.

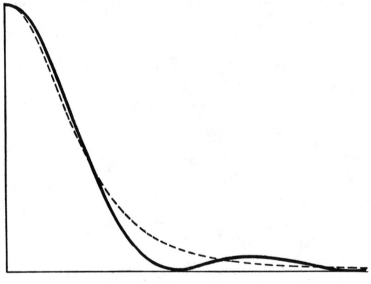

FIG. 57

Consider a grating in which the phase change, 2β, occurs at the middle, that is, $\Phi = \cos(\kappa x - \beta)$ for the first half and $\Phi = \cos(\kappa x + \beta)$ for the last half. Then $uC = \sin\beta - \sin(\beta - ul)$, and $uS = 0$. This gives for $I = C^2$ the same value as for a "plane" surface one-half of which is lower than the other by the amount $\epsilon = \dfrac{\lambda}{2\pi}\beta$. The intensity of the diffraction figure is represented in Figure 58 for $\beta = 0$, $\pi/4$, $\pi/2$, and π.

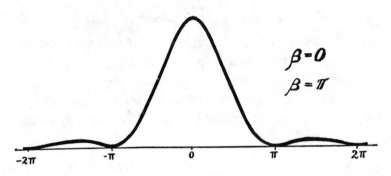

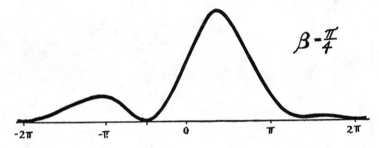

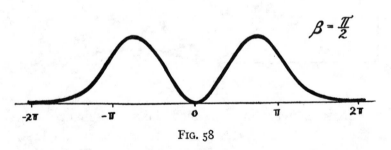

FIG. 58

As a third illustration, consider the case of a uniformly increasing grating space corresponding to

$$\sigma = \sigma_0 + cx ,$$

whence approximately

$$\beta = 2\pi c x^2 / \sigma^2 ,$$

and the result is the same as that of a plane wave with this phase error, namely, an alteration in focus, the amount of which is $\delta f = 2c\lambda f^2/\sigma^2$.

In general, any error in the ruling of a diffraction grating may be compensated by an appropriate change in phase, such as could be produced in an interposed plane-parallel plate by etching in dilute hydrofluoric acid.

GHOSTS

In consequence of the method of ruling (by means of a screw), the ruling will always have a periodic error, the period being the step of the screw.[1]
Let

$$\phi(x) = \cos m\kappa x \Sigma a_n \cos nqx \qquad \kappa = 2\pi/\sigma$$
$$+ \sin m\kappa x \Sigma b_n \sin nqx \qquad q = 2\pi/s \ .$$

The nth element of the summation is

$$\phi_n(x) = \cos m\kappa x + a_n \cos nqx \cos m\kappa x$$
$$+ b_n \sin nqx \sin m\kappa x \ .$$

This value substituted in

$$C = \int_{-l/2}^{l/2} \Phi(x) \cos uxdx \qquad (S = 0)$$

gives for the intensity of the nth ghost,

for $\quad \pm u = m\kappa \quad$ or $\sin\theta = m\lambda/\sigma \qquad I = 1$

for $\quad \pm u = m\kappa + nq$ or $\sin\theta = m\lambda/\sigma + n\lambda/s \quad I = (a_n + b_n)^2$

and for $\pm u = m\kappa - nq$ or $\sin\theta = m\lambda/\sigma - n\lambda/s \quad I = (a_n - b_n)^2$.

The distance between ghosts, $\delta\theta$, is $\dfrac{\lambda}{S\cos\theta}$.

---

[1] Similar effects arise from any other periodic change in the transmission, such as gears, belting, etc.

Let it be required to find the intensity of the nth ghost relative to that of the main line for a grating the nth element of which is

$$\phi(x) = \cos 2\pi \left(\frac{mx}{\sigma} - \frac{m\epsilon_n}{\sigma} \sin nqx \right)$$

in which ϵ_n/σ is small. This gives

$$a_0 = 1 \qquad\qquad b_0 = 0$$

$$a_n = 0 \qquad\qquad b_n = 2\pi m \frac{\epsilon_n}{\sigma},$$

whence

$$\frac{I_n}{I_0} = \left(2\pi m \frac{\epsilon_n}{\sigma} \right)^2.$$

CHAPTER IX

THE RULING OF DIFFRACTION GRATINGS

As shown in the preceding chapter, the maximum resolving power of a grating is

$$\overline{R} = 2l/\lambda\,,$$

that is, twice the number of light-waves in the entire length of the ruled surface. This shows that neither the closeness of the rulings nor the total number of lines[1] determines this theoretical limit. This can be reached, however, only on the condition of an extraordinary degree of accuracy in the spacing of the lines. Several methods for securing this degree of accuracy have been tried but none has proved as effective as the screw. This must be of uniform pitch throughout, and the periodic errors must be extremely small. For a short screw—for example, one sufficient for a grating 2 inches in length—the problem is not very difficult; but as the length of the screw increases, the difficulty increases much more rapidly. It was solved by Rowland in something over two years. Since this time many problems have arisen which demand a higher resolving power than even these gratings could furnish.[2]

[1] Practically, however, since a small number of lines requires a correspondingly high order of spectrum, and therefore, in general, a low intensity results, it is better to have a great number of lines with fairly close ruling, e.g., 500 to 1,000 lines per mm.

[2] Rowland's original gratings with about 100,000 lines had a resolving power of 150,000. The 6½-inch gratings now ruled on the Rowland engine have a much higher resolving power, perhaps 400,000.

Among these may be mentioned (1) the resolution of doublets and groups of lines whose complexity was unsuspected until revealed by the methods described in chapter iv; (2) the distribution of intensity within spectral "lines"; (3) their broadening and displacement with temperature and pressure; (4) the effect of magnetic and electric field; (5) the displacement of lines in consequence of motion of the source; (6) the separation of local disturbances in the solar atmosphere at different levels as shown in Hale's work with the spectroheliograph; (7) to these may now be added the exact measurement of the distribution of lines in "series" spectra, and the testing of the various theories which have been proposed to account for such distribution.

In the hope of increasing the resolving power of the diffraction grating to an extent which should materially assist in the investigation of such problems, the construction of a ruling engine was undertaken which could furnish gratings with a ruled space of 14 inches for which a screw of something over 20 inches is necessary. This screw was cut in a specially corrected lathe so that the original errors were not very large; and these were reduced by long grinding with very fine carborundum until it was judged that the residual errors were sufficiently small to be automatically corrected during the process of ruling.

The principal claim to novelty of treatment of the problem lies in the application of interference methods to the measurement and correction of these errors. For this purpose, one of the interferometer mirrors, A (Fig. 59), is fixed to the grating carriage, while a standard BC, consisting of two mirrors at a fixed distance apart, is attached to an auxiliary carriage. When the adjustment is correct for the surface B, interference fringes appear. The grating carriage is now moved through the length of the standard (one-tenth of a millimeter if the periodic error is under investigation; 10 mm or more if the error of run is to be determined) when the interference fringes appear on the surface C. This operation is repeated, and the difference from exact coincidence of the central (achromatic) fringe with a fiducial mark is measured at each step.[1] The corresponding correction is applied to the worm wheel by a simple automatic device which actuates

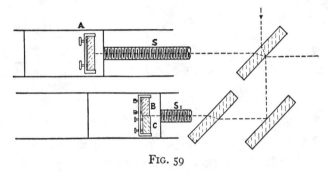

FIG. 59

the screw (for periodic errors) or to the nut which moves the grating carriage. In this way the residual errors have been almost completely eliminated and the resulting gratings have very nearly realized their theoretical efficiency.

A number of minor points may be mentioned which have contributed to the success of the undertaking: (a) The ways which guide the grating carriage as well as those which control the motion of the ruling diamond must be very true.[2] (b) The friction of the grating carriage

---

[1] This difference can be obtained easily to within one-twentieth of the fringe width, or to within a millionth of an inch.

[2] These were corrected by an auto-collimating device, the residual errors corresponding to deviations from a straight line of less than one second of arc.

on the ways was diminished to about one-tenth of that due to the weight (of grating and carriage) by floating on mercury. (*c*) The longitudinal motion of the screw was prevented by resting the spherical end against an optically plane surface of diamond, adjusted normal to the axis of the screw.

To test the performance of the grating it is set up in the Littrow mounting as shown in Figure 60. Light from the source (usually a Cooper-Hewitt lamp) is directed to the slit *s* at the focus of an excellent 8-inch achromatic

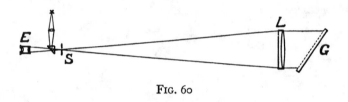

FIG. 60

lens *L* whence it proceeds as a parallel beam to the grating *G*. The light diffracted from the grating retraces its path, with a slight vertical displacement, forming an image of the slit immediately below the latter where it may be photographed or observed by the eyepiece *E*.

Figure 61 (facing p. 108) shows the spectrum of the green mercury line, λ 5461, given by a 10-inch grating in the sixth order. The resolving power, as shown by the reproduction in which the "width" of the sharper lines is about one and one-third divisions of the scale (one division equals 0.01 A.U.), is about 400,000. The original negative gives 600,000. The theoretical value is 660,000.

Doubtless the possibility of ruling a "perfect" grating by means of a homogeneous source has occurred to many; indeed this was one of the methods first attempted. It may still prove feasible if present attempts to produce still larger gratings meet with serious difficulty. Such a procedure may be made automatic, and would be independent of the mechanism.

An even simpler and more direct application of the light-waves from a homogeneous source is theoretically possible and perhaps experimentally practicable. If a point source of such radiation lie in the focus of a collimating lens, and the resulting plane wave be reflected at normal incidence from a true plane, stationary waves will be set up as in the Lippman plates; these will impress an inclined photographic plate with parallel bands, thus constituting a grating whose resolving power depends only on the degree of homogeneity of the source.

CHAPTER X

THE ECHELON GRATING

The attempt to rule gratings which throw all or most of the diffracted light into a single spectrum has had only a partial and more or less accidental success; and as the ruling diamond wears away even in the ruling of a single grating, the results cannot be relied upon in a repetition.[1]

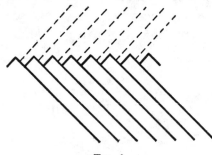

FIG. 62

The desired result may be obtained in another way as follows: If a pile of very accurate plane-parallel plates of exactly equal thickness be arranged as shown in Figure 62, and light is incident normal to the surfaces as shown in the dotted lines, the light will be reflected (and also diffracted) normally for such a wave-length that the constant difference in path between successive pencils is an exact whole number of light-waves. The resulting

[1] For very coarse gratings such as the "echelette" of R. W. Wood, in which the grating space is of the order of several hundredths of a millimeter, the results may be much more satisfactory.

spectrum will show light of this wave-length only at this particular angle for which the retardation $m\lambda = 2t$.

The difficulty in realizing such a reflecting "echelon" is that of insuring the exact equality of the air space between the plates.[1] This difficulty does not apply, however, to the light transmitted by such a system, the retardation in this case being $m\lambda = (\mu - 1)t$. ($m =$ an integer, $\mu =$ index of refraction, $t =$ thickness of the plates.)

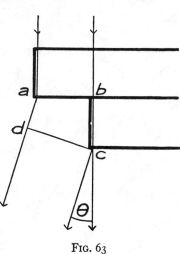

FIG. 63

If Figure 63 represents two plates of the echelon, the thickness $bc = t$, and the offset $ab = s$, then the retardation in direction θ will be $m\lambda = \mu bc - ad$.

$$m\lambda = \mu t - t \cos \theta + s \sin \theta \, ,$$

or, for small angles,

$$m\lambda = (\mu - 1)t + s\theta \, . \tag{1}$$

Differentiating for λ ($m =$ const.) gives the dispersion

$$\frac{d\theta}{d\lambda} = \frac{1}{s}\left(m - t\frac{d\mu}{d\lambda}\right) \, ,$$

or, since $m \doteq (\mu - 1)t/\lambda$, and putting $b = (\mu - 1) - \lambda\frac{d\mu}{d\lambda}$,

$$\frac{d\theta}{d\lambda} = \frac{bt}{s\lambda} \, . \tag{2}$$

[1] It is possible that sufficient accuracy might be obtained by "optical contact."

For constant λ

$$\frac{\Delta\theta}{\Delta m} = \frac{\lambda}{s} \, ,$$

or, if $\Delta m = 1$, the angle between successive orders of spectra will be

$$\Delta\theta = \lambda/s \, . \tag{3}$$

The limit of resolution

$$\partial\theta = \frac{\lambda}{a} = \frac{\lambda}{ns} \tag{4}$$

if n = number of plates. Substituting in (2) gives

$$\frac{\partial\lambda}{\lambda} = \frac{\lambda}{bnt} \, . \tag{4a}$$

Comparison of (3) and (4) shows that the limit of resolution is $1/n$th of the separation of consecutive orders.

If $\Delta\lambda$ is the difference in wave-length corresponding to successive orders, and $d\lambda$ is the required difference in wave-length of a spectral line from that of a given line, then

$$d\lambda = \frac{\Delta\lambda}{\Delta\theta} d\theta \, .$$

The value of the factor $\dfrac{\Delta\lambda}{\Delta\theta}$ may be found experimentally by counting the number of orders between known solar or iron lines.

The expression for the intensity

$$I = \frac{\sin^2 \pi s\theta/\lambda}{(\pi s\theta/\lambda)^2} \, ,$$

the graph of which is shown in Figure 64, gives $\theta_0 = \lambda/s$ for $I = 0$. This is also the angular distance between spectra.

There will be only two spectra $p_1 p_2$, excepting the case in which p_1 falls at $\theta = 0$, when there will be only one.[1]

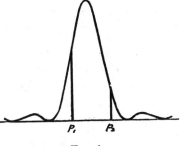

Fig. 64

Evidently the echelon spectroscope can only be used when the width of the spectral line is less than that which corresponds to $\Delta\theta = \lambda/s$, otherwise there will be over-lapping of successive spectra. This is also an inconvenience in the case of complicated groups such as occur in the spectra of mercury, or in the band spectra of nitrogen,

Fig. 65

etc., which may be eliminated by crossing the echelon by a grating or another echelon. Such an arrangement is shown in Figure 65, and Figures 66 and 67 show the spectrum of the sun and of the iron arc taken with this arrangement. Figure 68 shows the spectrum of the green mercury line.

[1] Neglecting the very faint spectra which occur beyond $\theta = \lambda/s$.

Two other instruments have been devised employing essentially the same principle as the echelon, namely, the Fabry and Perot apparatus and the Lummer-Gehrcke plate. The former consists of two silvered glass plates, the silvered surfaces parallel and facing each other. The silver films are semi-transparent, and light transmitted or reflected from the combination shows concentric circles in monochromatic light. The Lummer-Gehrcke plate employs the repeated reflections from the two surfaces of a very accurately plane-parallel plate.

In all three instruments there is a constant phase retardation δ. If the factor of reflection (or transmission) is r, the resultant vibration will be

$$R = \cos nt + r \cos nt \cos \delta + r^2 \cos nt \cos 2\delta + \cdots$$
$$+ r \sin nt \sin \delta + r^2 \sin nt \sin 2\delta + \cdots$$

or, if C is the coefficient of $\cos nt$ and S that of $\sin nt$,

$$C + iS = \frac{r^n e^{in\delta} - 1}{r e^{i\delta} - 1}$$

$$C - iS = \frac{r^n e^{-in\delta} - 1}{r e^{-i\delta} - 1}.$$

If $I_0 = 1$ for $\delta = 0$, the resulting intensity is

$$I = \frac{1 + \dfrac{4r^n \sin^2 n\delta/2}{(1 - r^n)^2}}{1 + \dfrac{4r \sin^2 \delta/2}{(1 - r)^2}}.$$

The angle θ corresponding to any given value of I may be obtained by substituting for δ its value in terms of θ.

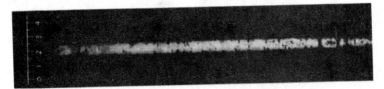

FIG. 66

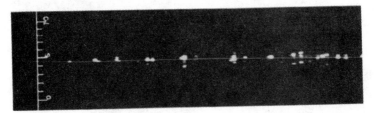

FIG. 67

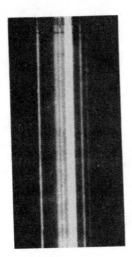

FIG. 61

FIG. 68

This for the echelon is given by

$$\frac{\delta}{2} = \frac{\pi d}{\lambda} = m\pi + \frac{\pi s \theta}{\lambda} \ ,$$

in which m is an integer

$$\sin \frac{\pi \delta}{2} = \sin \frac{\pi s \theta}{\lambda} \ ,$$

or if θ_1 be the angular separation of two successive orders of spectra, $\theta_1 = \frac{\lambda}{s}$, whence

$$\sin \frac{\pi \delta}{2} = \sin \frac{\pi \theta}{\theta_1} \ ,$$

and the formula for the echelon becomes

$$= I_\epsilon \frac{1 + \dfrac{4r^n}{(1-r^n)^2} \sin^2 n\pi \dfrac{\theta}{\theta_1}}{1 + \dfrac{4r}{(1-r)^2} \sin^2 \pi \dfrac{\theta}{\theta}} \ ,$$

in which $r =$ amplitude transmission factor.

For the Fabry-Perot instrument the retardation $d = 2t \cos \theta \doteq 2t - t\theta^2$, whence

$$\frac{\delta}{2} = \frac{\pi t}{\lambda} \left[2 - \theta^2 \right] \ ,$$

or, if $\frac{2t}{\lambda} = m$ (integer),

$$\sin \frac{\delta}{2} = \sin \frac{\pi t}{\lambda} \theta^2 \ .$$

If θ_1 is the angle corresponding to a difference of path λ,
$\theta_1^2 = \dfrac{\lambda}{t}$, whence

$$I_1 = \frac{1 + \dfrac{4r^{2n}}{(1-r^{2n})^2}\sin^2 2\pi n \dfrac{\theta^2}{\theta_1^2}}{1 + \dfrac{4r^2}{(1-r^2)^2}\sin^2 2\pi \dfrac{\theta^2}{\theta_1^2}},$$

in which $r =$ amplitude reflection factor.

An echelon of fourteen elements for which $r = .99$ gave for limit of resolution $\dfrac{\delta\theta}{\theta_1} \doteqdot \dfrac{1}{15}$ agreeing fairly well with calculation.[1]

A Perot and Fabry apparatus for which $r^2 = .9$ gave

$$\frac{\delta\theta}{\theta_1} = \frac{1}{10}.$$

[1] It may be possible to construct a reflecting echelon by silvering the surfaces of a transmission echelon whose surfaces are in optical contact. For such an echelon $r = 1$ and the expression for the limit of resolution reduces to $\dfrac{\delta\theta}{\theta_1} = \dfrac{1}{n}$.

The limit, in terms of λ, is given by $\dfrac{\delta\lambda}{\lambda} = \dfrac{\lambda}{2l}$, which is the same as the extreme value for a grating of equal length under grazing incidence.

CHAPTER XI

APPLICATION OF INTERFERENCE TO ASTRONOMY

The preceding diffraction formulae apply to the case of a single point source. If the object presents a number of such points, the total effect at the focus would be given by the summation of the individual intensities[1]

$$I = \Sigma_n(C^2 + S^2) ,$$

in which, if u_n and v_n are the co-ordinates of the nth point,

$$C = \frac{1}{\lambda^2 f^2} \int \int \Phi \cos [(u-u_n)x+(v-v_n)y+\psi]dxdy$$

$$S = \frac{1}{\lambda^2 f^2} \int \int \Phi \sin [(u-u_n)x+(v-v_n)y+\psi]dxdy .$$

In the case of constant Φ and ψ for points sufficiently near to $v=0$, omitting the constant factor,

$$C = \int \cos (u-u_n)x \, dx$$

$$S = \int \sin (u-u_n)x \, dx .$$

In the particular case of two equal rectangular apertures,

$$I = \Sigma A^2 \cos^2 \frac{\pi}{\lambda} s(a-a_n)$$

$$A = \frac{\sin \frac{\pi}{\lambda} a(a-a_n)}{\frac{\pi}{\lambda} a(a-a_n)} ,$$

[1] The elements of sources are luminous points which have no constant phase relation.

in which a is the width of the apertures and s the distance apart of their centers. If a is small compared with s, A will be approximately equal to unity.

For a continuous group the summation becomes an integration. Replacing a_n by a and a by β, and putting $\phi(a)da$ for the intensity of a strip of the object of width da,

$$I = \int \phi(a) \cos^2 \frac{\pi}{\lambda} s(\beta - a)\, da$$

or if

$$2\frac{\pi}{\lambda} s\beta = \theta \qquad C = \int \phi(a) \cos ka\, da$$

$$S = \int \phi(a) \sin ka\, da \qquad P = \int \phi(a)da \qquad k = 2\pi s/\lambda ,$$

$$I = P + C \cos \theta + S \sin \theta ,$$

whence, as shown in the similar expression (chap. iv, p. 36), the visibility V of the interference fringes is given by

$$P^2 V^2 = C^2 + S^2 .$$

DOUBLE STAR

For a double star the brightness of whose components is in the ratio of $1:r$,

$$V^2 = \frac{1 + r^2 + 2r \cos ka}{1 + r^2 + 2r} .$$

This expression is a minimum for

$$ka = n\pi \qquad (n = 1, 3, 5 \cdots) ,$$

or for

$$a = \frac{n\lambda}{2s} .$$

Accordingly, this angle can be accurately measured even when less than half the limit of resolution of the full aperture of the observing telescope.

Again, by comparing the visibility at maximum and at minimum, the ratio of the brightness of the component stars may be found by

$$r = \frac{V_{max.} - V_{min.}}{V_{max.} + V_{min.}} .$$

STAR DISKS

For a uniformly[1] illuminated disk,

$$V = \int_0^1 \sqrt{1 - \omega^2} \cos^n \omega d\omega ,$$

in which

$$n = \pi a \frac{s}{\lambda} ,$$

a being the angular diameter. For such an object the fringes vanish for $a = 1.22\lambda/s$.

The feasibility of the interference method for measurement of luminous disks was amply confirmed in an investigation with the 12-inch telescope at Mount Hamilton, in which the diameters of the satellites of Jupiter

[1] If the illumination can be represented as a function of the distance from the center $I = (R^2 - r^2)^n$ then it can be shown that the visibility of the fringes will be given by

$$V = \frac{\int_0^R (R^2 - x^2)^{\frac{2n+1}{2}} \cos kx dx}{\int_0^R (R^2 - x^2)^{\frac{2n+1}{2}} dx} .$$

[Footnote continued on page 114]

were measured with an order of accuracy considerably greater than could be obtained by the usual method. The step from such diameters (of the order of one second) to that of the stars, the largest and nearest of which were surmised to subtend an angle of a few hundredths of a second, was so considerable that it might well be questioned whether atmospheric and mechanical disturbances might not prevent realization of the undertaking.

The graphs for $n=0, \frac{1}{2}, 1, 2$, as calculated by Professor F. R. Moulton, are shown in Fig. 69. If future observations may be made with the requi-

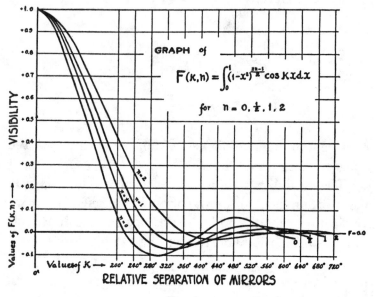

FIG. 69

site accuracy of the distances $b_1 b_2$, at which the fringes vanish for the first and second times, a fair approximation to the value of n in the light-curve of the star will be given by

$$n = \frac{5.5\, b_1 - 3 b_2}{b_2 - b_1}.$$

It is doubtful if a telescope could be constructed large enough to observe the vanishing of the interference of the fringes due to the double aperture in front of the objective; but while such a telescope would be practically impossible, the interferometer (utilizing two plane mirrors instead of the two apertures) may be employed, in which there is no limit to the effective base line except that imposed by such atmospheric and mechanical disturbances.

In order to test the effect of atmospheric disturbances, trials with the 40-inch Yerkes refractor and the 60-inch and the 100-inch reflectors at Mount Wilson were made which, even with relatively poor "seeing," gave excellent results, showing clearly that up to these distances and inferentially to much greater, the effect of atmospheric disturbances were not to be feared.[1]

The limitation of the incident light to two pencils can be effected much more readily and effectively by a relatively small diaphragm with two apertures not far from the focus of the telescope than by a screen covering the objective. Such a small diaphragm was used in the observations with the 60-inch and the 100-inch tele-

[1] Doubtless this rather unexpected result may be explained as follows: The confusion of the image in poor seeing is due to the integrated effect of elements of the incident light-waves, elements which are not in constant phase relation in consequence of inequalities in the atmosphere due to temperature differences; the optical result being a "boiling" of the image, closely resembling the appearances of objects viewed over a heated surface.

In the case of the two elements at opposite ends of a diameter of the objective, the same differences in phase produce a motion of the (straight) interference fringes (and not a confusion) and if, as is usually the case, this motion is not too rapid for the eye to follow, the visibility of the fringes is quite as good as in the case of perfect atmospheric conditions.

scopes; and it was intended to attempt measurements of double stars by such an arrangement with the addition of a means of measuring the variable distance between the apertures at which the fringes disappeared.

Dr. Anderson, of the staff of the Mount Wilson Observatory, preferred to rotate the diaphragm, keeping the distance between apertures constant. By this method the angular distance between the components of the double star, Capella, was measured. This star was known to be a doublet from spectroscopic observations of the Doppler displacement of the spectral lines; but the direct telescopic observation had never been made, although the calculated separation (about 0".05) would indicate the possibility if atmospheric conditions permitted. The resulting orbit for the components of Capella was measured with such extraordinary accuracy by the interference method that the calculated and observed positions agreed to one ten-thousandth of a second of arc.

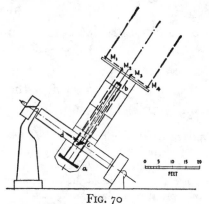

FIG. 70

The trials with the 60-inch and with the 100-inch telescopes having shown that up to these distances the effect of atmospheric disturbances were not to be feared, it was decided to build an interferometer with movable mirrors (Fig. 70), permitting tests to be made with a base as great as 20 feet. The interferometer beam (Figs. 71 and 72) was made of structural steel as rigid as possible, with the minimum admissible weight. (It was found that this could not be reduced to less than 800 pounds—a truly formidable addition to any but the most ruggedly constructed instrument such as the 100-inch. It may be noted that it was for this reason that the 100-inch telescope was utilized to support the interferometer and not at all on account of its optical power.) As shown in the

FIG. 71

diagram (Fig. 70), the light from a star is received on the outer movable mirrors M_1M_4, whence they are reflected to the fixed inner mirrors M_2M_3, proceeding thence to the concave reflector a, to the convex mirror b, and thence to the inclined plane mirror c, whence they are brought together at d where the interference fringes are observed by means of a low-power eyepiece. Coincidence of the two interfering pencils is facilitated by the tilting of interposed plane-parallel glass plates; and equalization of the two paths adjusted by the help of a double wedge of glass. With this construction, interference fringes were readily

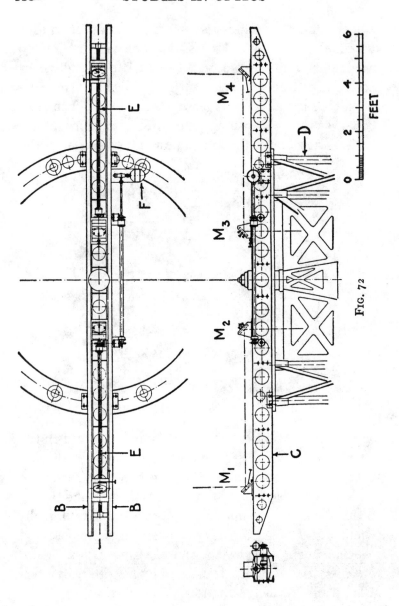

FIG. 72

observed at the extreme distance of 20 feet, showing that even at this large separation neither atmospheric conditions nor vibrations and strains would seriously interfere with observations.

The first actual observation which furnished measurable results of star diameter was made by Mr. F. G. Pease, of the staff of the Mount Wilson Observatory. The instrument was pointed at the red giant star Betelgeuse in the constellation of Orion; and when the separation of the outer mirrors of the interferometer was 10 feet, the interference fringes vanished, although the fringes were readily observed at this separation when the instrument was directed to other stars (β Persei and γ Orionis).

Assuming the effective wave-length of the light from Betelgeuse at 5.75×10^{-5} cm and with $s = 121$ in. (306.5 cm) the angular diameter of this star $a = 1.22\lambda/s$ proves to be 0".047. An estimate of its linear diameter using a parallax 0".018 gives 240×10^6 miles (about the diameter of the orbit of Mars).

This calculation is based on the supposition of a uniformly illuminated disk, which is doubtless far from the fact. A darkening toward the limb equal to that of the sun would result in an increase of the estimated diameter of about 17 per cent. Since this observation, other stars, some of which are even larger (Arcturus, Mira, Antares), have been measured.

A still larger interferometer is now under construction with a base of 50 feet which should increase the power of the instrument about two and one-half times.

CHAPTER XII

VELOCITY OF LIGHT

The velocity of light is one of the most important of the fundamental constants of Nature. Its measurement by Foucault and Fizeau gave as the result a speed greater in air than in water, thus deciding in favor of the undulatory and against the corpuscular theory. Again, the comparison of the electrostatic and the electromagnetic units gives as an experimental result a value remarkably close to the velocity of light—a result which justified Maxwell in concluding that light is the propagation of an electromagnetic disturbance. Finally, the principle of relativity gives the velocity of light a still greater importance, since one of its fundamental postulates is the constancy of this velocity under all possible conditions.

The first attempt at measurement was due to Galileo. Two observers, placed at a distance of several kilometers, are provided with lanterns which can be covered or uncovered by a movable screen. The first observer uncovers his light, and the second observer answers by uncovering his at the instant of perceiving the light from the first. If there is an interval between the uncovering of the lantern by the first observer and his perception of the return signal from the second (due allowance being made for the delay between perception and motion), the distance divided by the time interval should give the velocity of propagation.

Needless to say, the time interval was far too small to be appreciated by such imperfect appliances. It is nevertheless worthy of note that the principle of the method is sound, and, with improvements that are almost intuitive, leads to the well-known method of Fizeau. The first improvement would clearly be the substitution of a mirror instead of the second observer. The second would consist in the substitution of a series of equidistant apertures in a rapidly revolving screen instead of the single screen which covers and uncovers the light.

The first actual determination of the velocity of light was made in 1675 by Römer as a result of his observation of the eclipses of the first satellite of Jupiter. These eclipses, recurring at very nearly equal intervals, could be calculated, and Römer found that the observed and the calculated values showed an annual discrepancy. The eclipses were later by an interval of sixteen minutes and twenty-six seconds[1] when the earth is farthest from Jupiter than when nearest to it. Römer correctly attributed this difference to the time required by light to traverse the earth's orbit. If this be taken as 300,000,000 kilometers and the time interval as one thousand seconds, the resulting value for the velocity of light is 300,000 kilometers per second.

Another method for the determination of the velocity of light is due to Bradley, who in 1728 announced an apparent annual deviation in the direction of the fixed stars from their mean position, to which he gave the name "aberration." A star whose direction is at right angles to the earth's orbital motion appears displaced in the direc-

[1] The value originally given by Römer, twenty-two minutes, is clearly too great.

tion of motion by an angle of 20″.445. This displacement
Bradley attributed to the finite velocity of light.

With a telescope pointing in the true direction of such
a star, during the time of passage of the light from objective to focus the telescope will have been displaced in
consequence of the orbital motion of the earth so that the
image of the star falls behind the crosshairs. In order to
produce coincidence, the telescope must be inclined for-
ward at such an angle a that the tangent is equal to the
ratio of the velocity v of the earth to the velocity of light,

$$\tan a = \frac{v}{V} ,$$

or, since $v = \pi D/T$, where D is the diameter of the earth's
orbit and T the number of seconds in the year,

$$\tan a = \frac{\pi D}{VT} ,$$

from which the velocity of light may be found; but, as is
also the case with the method of Römer, only to the degree of accuracy with which the sun's distance, $\frac{1}{2}D$, is
known; that is, with an order of accuracy of about 1 per
cent.[1]

In 1849 Fizeau announced the result of the first experimental measurement of the velocity of light. Two
astronomical telescope objectives L_1 and L_2 (Fig. 73) are
placed facing each other at the two stations. At the focus
of the first is an intense but minute image a of the source
of light (arc) by reflection from a plane-parallel plate N.

---

[1] The value of the velocity of light has been obtained, by experimental
methods immediately to be described, with an order of accuracy of one in
one hundred thousand, so that now the process is inverted, and this result is employed to find the sun's distance.

The light from this image is rendered approximately parallel by the first objective. These parallel rays, falling on the distant objective, are brought to a focus at the surface of a mirror, whence the path is retraced and an image formed which coincides with the original image a, where it is observed by the ocular E. An accurately

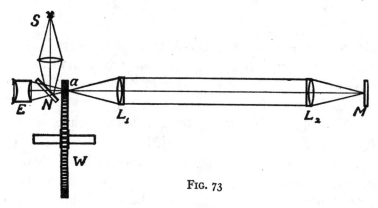

Fig. 73

divided toothed wheel W is given a uniform rotation, thus interrupting the passage of the light at a. If, on returning, the light is blocked by a tooth, it is eclipsed, to reappear at a velocity such that the next succeeding interval occupies the place of the former, and so on.

If n is the number of teeth and N the number of turns per second, K the number of teeth which pass during the double journey of the light over the distance D,

$$V = \frac{2NnD}{K} .$$

It is easier to mark the minima than the maxima of intensity, and accordingly

$$K = \frac{2p-1}{2}$$

if p is the order of the eclipse. Let δK be the error committed in the estimate of K (practically the error in estimation of equality of intensities on the descending and the ascending branches of the intensity curve). Then

$$\frac{\partial V}{V} = \frac{\partial K}{K} .$$

Hence it is desirable to make K as great as possible. In Fizeau's experiments this number was 5 to 7, and should have given a result correct to about one three-hundredth. It was, in fact, about 5 per cent too large.

A much more accurate determination was undertaken by Cornu in 1872 in which K varied from 3 to 21, the result as given by Cornu being 300,400, with a probable error of one-tenth of 1 per cent. In discussing Cornu's results, however, Listing showed that these tended toward a smaller value as the speed increased, and he assigns this limit as the correct value, namely, 299,950. Perrotin, with the same apparatus, found 299,900.

Before Fizeau had concluded his experiments, another project was proposed by Arago, namely, the utilization of the revolving mirror by means of which Wheatstone had measured the speed of propagation of an electric current. Arago's chief interest in the problem lay in the possibility of deciding the question of the relative velocities in air and water as a crucial test between the undulatory and the corpuscular theories. He pointed out, however, the possibility of measuring the absolute velocity.

The plan was to compare the deviations of the light from an electric spark reflected directly from the revolving mirror with that which was reflected after traversing a considerable distance in air (or in water). The difficulty in executing such an experiment lay in the uncertainty in the direction in which the two reflected images of the spark were to appear (which might be anywhere in 360°). This difficulty was solved by Foucault in 1862 by the following ingenious device whereby the return light is

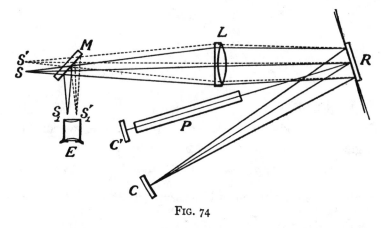

FIG. 74

always reflected in the same direction (apart from the deviation due to the retardation which it is required to measure), notwithstanding the rotation of the mirror.

Following is the actual arrangement of apparatus by which this is effected. Light from a source S falls upon an objective L, whence it proceeds to the revolving mirror R, and is thence reflected to the concave mirror C (whose center is at R), where it forms a real image of the source. It then retraces its path, forming a real image which coincides with the source even when the revolving mirror is in slow motion. Part of the light is reflected from the plane-parallel glass M, forming an image at a where it is observed by the micrometer eyepiece E.

If now the revolving mirror is turning rapidly, the return image, instead of coinciding with its original position, will be deviated in the direction of rotation through an angle double that through which the mirror turns while the light makes its double transit. If this angle is a and the distance between mirrors is D, and the revolving mirror makes N turns per second,

$$a = 2\pi N \frac{2D}{V},$$

or

$$V = \frac{4\pi N D}{a}.$$

In principle there is no essential difference between the two methods. In the method of the toothed wheel the angle a corresponds to the passage of K teeth, and is therefore $a = 2\pi K/n$, so that the formula previously found, $V = \frac{2NnD}{K}$, now becomes $V = \frac{4\pi ND}{a}$, the same as for the revolving mirror. The latter method has, however, the same advantage over the former that the method of mirror and scale has over the direct reading of the needle of a galvanometer.

On the other hand, an important advantage for the method of the toothed wheel lies in the circumstance that the intensity of the return image is one-half of that which would appear if there were no toothed wheel, whereas with the revolving mirror this fraction is $\frac{n\beta}{rD}$ if the mirror has n facets), where β is the angular aperture of the concave mirror, and f is the focal length of the mirror, r is the distance from slit to revolving mirror, and D is the distance between stations.

In the actual experiments of Foucault, the greatest distance D was only 20 m (obtained by five reflections from concave mirrors), which, with a speed of five hundred turns per second, gives only 160″ for the angle $2a$ which is to be measured. The limit of accuracy of the method is about one second, so that under these conditions the results of Foucault's measurements can hardly be expected to be accurate to one part in one hundred and sixty. Foucault's result, 298,000, is in fact too small by this amount.[1]

In order to obtain a deflection $2a$ sufficiently large to measure with precision it is necessary to work with a much larger distance. The following plan renders this possible, and in a series of experiments (1878) the distance D was about 700 m and could have been made much greater.

The image-forming lens in the new arrangement is placed between the two mirrors, and (for maximum intensity of the return image) at a distance from the revolving mirror equal to the focal length of the lens. This necessitates a lens of long focus; for the radius of measurement r (from which a is determined by the relation $\delta = r \tan a$, in which δ is the measured displacement of the image) is given by $r = \dfrac{f^2}{D}$, if f is the focal length of the lens; whence r is proportional to f^2. In the actual experiment,

---

[1] Apart from the mere matter of convenience in limiting the distance D to the insignificant 20 m (on account of the dimensions of the laboratory), it may be that this was in fact limited by the relative intensity of the return image as compared with that of the streak of light caused by the direct reflection from the revolving mirror, which in Foucault's experiments was doubtless superposed on the former. The intensity of the return image varies inversely as the cube of the distance, while that of the streak remains constant.

a non-achromatic lens of 25-m focus and 20-cm diameter was employed, and with this it was found that the intensity of the return light was quite sufficient even when the revolving mirror was far removed from the principal focus.

With so large a displacement, the inclined plane-parallel plate in the Foucault arrangement may be suppressed, the direct (real) image being observed. With 250 to 300 turns per second, a displacement of 100 to 150 mm was obtained which could be measured with an error of less than one ten-thousandth.

The measurement of D presents no serious difficulty. This was accomplished by means of a steel tape whose coefficient of stretch and of dilatation was carefully determined, and whose length under standard conditions was compared with a copy of the standard meter. The estimated probable error was of the order of 1:200,000.

The measurement of the speed of rotation presents some points of interest. The optical "beats" between the revolving mirror and an electrically maintained tuning fork were observed at the same time that the coincidence of the deflected image with the crosshairs of the eyepiece was maintained by hand regulation of an air blast which actuated the turbine attached to the revolving mirror. The number of vibrations of the fork *plus* the number of beats per second gives the number of revolutions per second in terms of the rate of the fork. This, however, cannot be relied upon except for a short interval, and it was compared before and after every measurement with a standard fork. This fork, whose temperature coefficient is well determined, is then compared, as follows, directly with a free pendulum.

For this purpose the pendulum is connected in series with a battery and the primary of an induction coil whose circuit is interrupted by means of a platinum knife edge attached to the pendulum passing through a globule of mercury. The secondary of the induction coil sends a flash through a vacuum tube, thus illuminating the edge of the fork and the crosshair of the observing microscope. If the fork makes an exact whole number (256) of vibrations during one swing of the pendulum, it appears at rest; but if there is a slight excess, the edge of the fork appears to execute a cycle of displacement at the rate of n per second. The rate of the fork is then $N \pm n$ per second of the free pendulum. This last is finally compared with a standard astronomical clock.[1] The order of accuracy is estimated as $1:200,000$.

The final result of the mean of two such determinations of the velocity of light made under somewhat similar conditions but at a different time and locality is 299,895.

A determination of the velocity of light by a modification of the Foucault arrangement was completed by Newcomb in 1882. One of the essential improvements consisted in the use of a revolving steel prism with square section twice as long as wide. This permits the sending and receiving of the light on different parts of the mirror, thus eliminating the effect of direct reflection. It should also be mentioned that very accurate means were provided for measuring the deflection, and finally that the speed of the mirror was registered on a chronograph through a system of gears connected with the revolving mirror. Newcomb's result is 299,860.

[1] The average beat of such a clock may be extremely constant although the individual "seconds" vary considerably.

The original purpose of the Foucault arrangement was the testing of the question of the relative velocities of light in air and in water. For this purpose a tube filled with water and closed with plane-parallel glasses is interposed. There are then two return images of the source which would be superposed if the velocities were the same. By appropriately placed diaphragms these two images may be separated, and if there is any difference in velocities this is revealed by a relative displacement in the direction of rotation. This was found greater for the beam which had passed through the water column, and in which, therefore, the velocity must have been less. This result is in accordance with the undulatory theory and opposed to the corpuscular theory of light.

The experiments of Foucault do not appear to have shown more than qualitative results, and it should be of interest, not only to show that the velocity of light is less in water than in air, but that the ratio of the velocities is equal to the index of refraction of the liquid. Experiments were accordingly undertaken with water, the result obtained agreeing very nearly with the index of refraction. But on replacing the water by carbon disulphide, the ratio of velocities obtained was 1.75 instead of 1.64, the index of refraction. The difference is much too great to be attributed to errors of experiment.

Lord Rayleigh found the following explanation of the discrepancy. In the method of the toothed wheel the disturbances are propagated in the form of isolated groups of wave-trains. Rayleigh finds that the velocity of a group is not the same as that of the separate waves except in a medium without dispersion. The simplest form of group analytically considered is that produced by two simple harmonic wave-trains of slightly different frequencies and wave-lengths. Thus, let

$$y = \cos(nt - mx) + \cos(n_1t - m_1x) ,$$

in which $n = 2\pi/T$, and $m = 2\pi/\lambda$, T being the period and λ the wave-length. Let $n - n_1 = \partial n$, and $m - m_1 = \partial m$. Then

$$y = 2 \cos \tfrac{1}{2}(\partial nt - \partial mx) \cos(nt - mx) .$$

This represents a series of groups of waves such as illustrated in Figure 75.

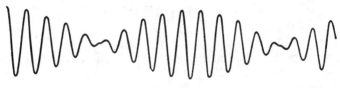

FIG. 75

The velocity of the waves is the ratio $V = n/m$, but the velocity of the group (e.g., the velocity of propagation of the maximum or the minimum) will be

$$V' = \partial n/\partial m ,$$

or, since $n = mV$,

$$V' = \frac{\partial(mV)}{\partial m} = V + m\frac{\partial V}{\partial m} = V\left(1 + \frac{m\partial V}{V\partial m}\right) ,$$

or, since $m = 2\pi/\lambda$,

$$V' = V\left(1 - \frac{\lambda}{V}\frac{\partial V}{\partial \lambda}\right) .$$

The demonstration is true, not only of this particular form of group, but (by the Fourier theorem) can be applied to a group of any form.

It is not quite so clear that this expression applies to the measurements made with the revolving mirror. Lord Rayleigh shows that in consequence of the Doppler effect there is a shortening of the waves at one edge of the beam of light reflected from the revolving mirror and a lengthening at the opposite edge, and since the velocity of propagation depends on the wave-length in a dispersive medium, there will be a rotation of the individual wavefronts.

If ω is the angular velocity of the mirror, and ω_1 that of the dispersional rotation,

$$\omega_1 = \frac{dV}{dy} = \frac{dV}{d\lambda}\frac{d\lambda}{dy},$$

where y is the distance from the axis of rotation. But

$$\frac{d\lambda}{dy} = 2\omega\frac{\lambda}{V} \therefore \omega_1 = 2\omega\frac{\lambda}{V}\frac{dV}{d\lambda}.$$

The deflection actually observed is therefore

$$T(2\omega+\omega_1),$$

where T is the time required to travel distance $2D$; or

$$\frac{4D}{V}\omega\left(1+\frac{\lambda}{V}\frac{dV}{d\lambda}\right),$$

hence the velocity measured is

$$V'' = V\div\left(1+\frac{\lambda}{V}\frac{dV}{d\lambda}\right),$$

or, to small quantities of the second order,

$$V'' = V' = \text{group velocity.}[1]$$

---

[1] J. W. Gibbs (*Nature*, 1886) shows that the measurement is in reality exactly that of groups and not merely an approximation.

The value of $\left(1+\dfrac{\lambda}{\mu}\dfrac{d\mu}{d\lambda}\right)$ for carbon disulphide for the mean wave-length of the visible spectrum is 0.93. Accordingly,

$$\frac{V_0}{V'}=\frac{V_0}{V}\frac{1}{0.93}=\frac{1.64}{0.93}=1.76 ,$$

which agrees with the value found by experiment.

RECENT MEASUREMENTS OF THE VELOCITY OF LIGHT

In the expression for V, the velocity of light as determined by the revolving mirror, $V=4\pi ND/a$, there are three quantities to be measured, namely, N, the speed of the mirror; D, the distance between stations; and a, the angular displacement of the mirror. As has already been mentioned, the values of N and D may be obtained to one part in one hundred thousand or less. But a cannot be measured to this order of accuracy. It has been pointed out by Newcomb[1] that this difficulty may be avoided by giving the revolving mirror a prismatic form and making the distance between the two stations so great that the return light is reflected at the same angle by the next following face of the prism.

The following is an outline of a proposed attempt to realize such a project between Mount Wilson and Mount San Antonio near Pasadena, the distance being about 35 km. For this, given a speed of rotation of 1,060 turns per second, the angular displacement of the mirror during the double journey would be 90°; or, if the speed were half as great, an angle of 45° would suffice.[2] Accordingly,

[1] *Measures of the Velocity of Light.* Nautical Almanac Office, 1882.

[2] It may be noted that with eight surfaces the resulting intensity will be four times as great as with the revolving plane-parallel disk.

the revolving mirror may have the form of an octagon. It is, of course, very important that the angles should be equal, at least to the order of accuracy desired.

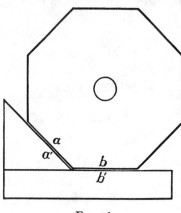

FIG. 76

This has already been attained as follows. The octagon, with faces polished and angles approximately correct, is applied to the test angle $a'b'$ made up of a 45° prism cemented to a true plane. The faces b_1b are made parallel by the interference fringes observed in monochromatic light. In general, the faces a_1a will not be parallel, and the angle between them is measured by the distance and inclination of the interference bands. The same process is repeated for each of the eight angles, and these are corrected by repolishing until the distance and inclination are the same for all, when the corresponding angles will also be equal. It has been found possible in this way to produce an octagon in which the average error was of the order of one-millionth, that is, about one-tenth to one-twentieth of a second.[1]

Another difficulty arises from the direct reflection and the scattered light from the revolving mirror. The former may be eliminated, as already mentioned, by slightly

[1] It may be noted that while a distortion may be expected when the mirror is in such rapid rotation, if the substance of the mirror (glass, in the present instance) is uniform, such distortion could only produce a very slight curvature and hence merely a minute change of focus.

inclining the revolving mirror, but to avoid the scattered light it is essential that the return ray be received on a different surface from the outgoing.

Again, in order to avoid the difficulty in maintaining the distant mirror perpendicular to the incident light, the return of the ray to the home station may be accom-

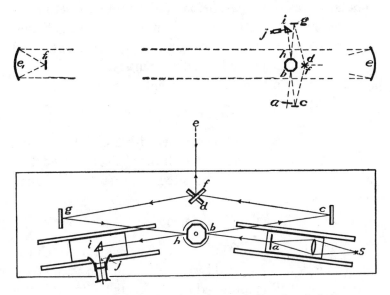

FIG. 77.—Light path a, b, c, d, e, e_1, f_1, l_1, e, f, g, h, i, j

plished exactly as in the Fizeau experiment, the only precaution required being the very accurate focusing of the beam on the small plane (better, concave) mirror at the focus of the distant collimator.

Finally, it is far less expensive to make both sending and receiving collimators silvered mirrors instead of lenses.

In Figure 77 is shown the arrangement of apparatus which fulfilled all these requirements.

Three determinations were undertaken between the home station at the Mount Wilson Observatory and Mount San Antonio 22 miles distant. The rate of the electric tuning fork was 132.25 vibrations per second, giving four stationary images of the revolving mirror when this was rotating at the rate of 529 turns per second. The fork was compared before and after every set of the observations with a free pendulum whose rate was found by comparison with an invar pendulum furnished and rated by the Coast and Geodetic Survey.

The result of eight measurements in 1924 gave

$$V_a = 299,735 .$$

Another series of observations with a direct comparison of the same electric fork with the Coast and Geodetic Survey pendulum[1] was completed in the summer of 1925 with a resulting value

$$V_a = 299,690 .$$

A third series of measurements was made in which the electric fork was replaced by a free fork making 528 vibrations per second maintained by an "audion circuit," thus insuring a much more nearly constant rate. The result of this measurement gave

$$V_a = 299,704 .$$

Giving these determinations the weights 1, 2, and 4, respectively, the result for the velocity in air is

$$V_a = 299,704 .$$

[1] This comparison was made by allowing the light from a very narrow slit to fall on a mirror attached to the pendulum. An image of the slit was formed by means of a good achromatic lens, in the plane of one edge of the fork, where it was observed by an ordinary eyepiece.

Applying the correction of 67 km for the reduction to *vacuo* gives finally $V = 299{,}771$.

This result should be considered as provisional, and depends on the value of D, the distance between the two stations which was furnished by the Coast and Geodetic Survey, and which it is hoped may be verified by a repetition of the work.

It was also found that a trial with a much larger revolving mirror gave better definition, more light, and steadier speed of rotation; so that it seems probable that results of much greater accuracy may be obtained in a future investigation.

FINAL MEASUREMENTS

Observations with the same layout were resumed in the summer of 1926, but with an assortment of revolving mirrors.

The first of these was the same small octagonal glass mirror used in the preceding work. The result obtained this year was $V = 299{,}813$. Giving this a weight 2 and the result of preceding work weight 1 gives 299,799 for the weighted mean.

The other mirrors were a steel octagon, a glass 12-sider, a steel 12-sider, and a glass 16-sider.

The final results are summarized in Table VII.

TABLE VII

Turns per Second	Mirror	Number of Observations	Vel. of Light in *Vacuo*
528	Glass oct.	576	299,797
528	Steel oct.	195	299,795
352	Glass 12	270	299,796
352	Steel 12	218	299,796
264	Glass 16	504	299,796

Weighted mean, $299{,}796 \pm 1$

Table VIII shows the more reliable results of measurements of V with distance between stations, method used, and the weight assigned to each.

TABLE VIII

Author	D	Method	Wt.	V
Cornu.........	23 km	Toothed wheel	1	299,990
Perrotin.......	12	Toothed wheel	1	299,900
M_1 and M_2.....	0.6	Rev. mirror	1	299,880
Newcomb*.....	6.5	Rev. mirror	3	299,810
M_3...........	35	Rev. mirror	5	299,800

* Newcomb's value omitting all discordant observations was 298,860.

CHAPTER XIII*

EFFECTS OF MOTION OF THE MEDIUM ON VELOCITY OF LIGHT

EFFECTS OF FIRST ORDER

The effects of motion of the medium on the velocity of light were first brought to the notice of the physicist in consequence of an experiment of Airy which had for its object the testing of a theoretical consequence of the undulatory theory regarding the problem of aberration. This requires that the tangent of the aberration shall be equal to the ratio of the velocity of the earth in its motion around the sun to the velocity of light. With a medium of index of refraction μ, this relation would be

$$\tan a = \mu \frac{v}{V},$$

and accordingly, if the observing telescope were filled with water the aberration should come out $4/3$ times as great as with air. The experiment was actually carried out by Airy with the result that the aberration was found to be the same as before.

The explanation of the difficulty, proposed by Fresnel, is based on the hypothesis that the luminiferous medium is carried along (*entrainé*) by the motion of the medium; not, however, by the full amount of this motion, but by a fraction, $\rho = \dfrac{\mu^2 - 1}{\mu^2}$, known as the "Fresnel coefficient." Taking the negative experimental result as basis, this value for ρ may be deduced as follows:

139

* See note, p. v.

Let *ac* (Fig. 78) be a plane wave-front from a star reaching the objective of the observing telescope which must be inclined at an angle *a* in consequence of the

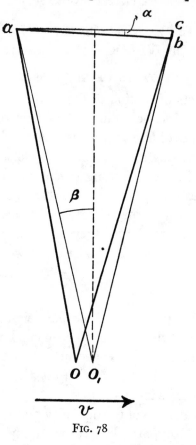

FIG. 78

motion of the earth in the direction of *v*. From the margins of the objective *a* and *c*, two rays arrive at o_1 (as do all the others) in the same phase. The time of passage of the two rays is then the same, whence

$$\frac{ao_1}{V_1} = \frac{cb}{V} + \frac{bo_1}{V_2},$$

in which

$$V_1 = V/\mu + \rho v \sin \beta$$

and

$$V_2 = V/\mu - \rho v \sin \beta$$

and β is the semi-aperture of the telescope. Substituting

$$ao_1 = ao + oo_1 \sin \beta$$
$$bo_1 = ao - oo_1 \sin \beta ,$$

and noting that

$$\frac{oo_1}{ao} = \mu \frac{v}{V} = \mu a ,$$

we find

$$2\mu a \sin \beta (1 - \rho) = \frac{bc}{\mu \cdot ao} .$$

But $2ao_1 \sin \beta$ is ac, the diameter of the objective, whence

$$\mu a(1 - \rho) = \frac{bc}{\mu \cdot ac} = \frac{a}{\mu} ,$$

and

$$\rho = \frac{\mu^2 - 1}{\mu^2} .$$

On the hypothesis of a stationary ether it appeared to be possible to detect a motion of the earth independent of astronomical observations, and many experiments of this kind were attempted, among which the following may be cited:

An interferometer is arranged as shown in Figure 79. In the path of the interfering pencils a glass prism of length D is interposed. Supposing the apparatus in motion in a direction parallel to the length D, the time of

traversing the circuit for the two pencils may be calculated as follows: During the passage of the light through

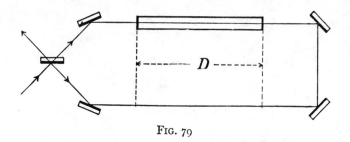

FIG. 79

the glass the apparatus will have moved through a
distance $\delta = \mu v/V(D+\delta)$, or with sufficient approximation,

$$\delta = \mu \frac{v}{V} D ,$$

while in the corresponding air path the apparatus will
have moved through

$$\delta_0 = \frac{v}{V} D .$$

Accordingly,

$$t_1 = \frac{D+\delta}{\dfrac{V}{\mu}+\rho v} + \frac{D-\delta_0}{V} ;$$

similarly, for the ray traversing the circuit in the opposite
sense,

$$t_2 = \frac{D-\delta}{\dfrac{V}{\mu}-\rho v} + \frac{D+\delta_0}{V}$$

As no displacement of the fringes is observed on rotating the whole apparatus through 180°, it follows that $t_1 = t_2$, whence

$$\frac{1+\mu\frac{v}{V}}{1+\mu\rho\frac{v}{V}} - \frac{1-\mu\frac{v}{V}}{1-\mu\rho\frac{v}{V}} = \frac{2\frac{v}{V}}{\mu},$$

or, to small quantities of second order,

$$1-\rho = 1/\mu^2 ,$$

or

$$\rho = \frac{\mu^2 - 1}{\mu^2} .$$

It may therefore be admitted that the Fresnel coefficient is a necessary consequence of the negative results of these endeavors to detect a motion of the earth relative to the luminiferous ether. It is desirable, none the less, to deduce this ratio from theoretical considerations.

This was first attempted by Eisenlohr as follows. Consider a prism of glass G (Fig. 80) of unit cross-section of index μ, in motion in the direction of

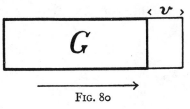

FIG. 80

the arrow with velocity v. Let the density of the ether outside be unity, and within the prism $1+\Delta$. Then in unit time the mass of ether introduced into the space v will be $m = v\Delta$. But on the entrainment hypothesis this may also be expressed by $m = \rho v(1+\Delta)$, whence

$$\Delta = \rho(1+\Delta) .$$

Since μ^2 is equal to the ratio of the densities of the ether within and without (the elasticities being supposed to remain constant),

$$\mu^2 = \frac{1+\Delta}{1} \; ,$$

and

$$\Delta = \mu^2 - 1 \; ,$$

which, substituted in the preceding expression, gives

$$\rho = \frac{\mu^2 - 1}{\mu^2} \; .$$

The following reasoning is based on the supposition that the ether outside the sphere of action of the molecules of matter is absolutely unaffected by the motion. Suppose a to be the diameter of the ether atmosphere carried along with a molecule in the direction of motion; and let b be the distance (average) between two molecules. While the light traverses the distance $a+b$, the system will have moved through a distance $a+\beta$. If ν is the index of refraction within the molecule,

$$a \doteqdot \nu \frac{v}{V} a$$

and

$$\beta = \frac{v}{V-v} b \; .$$

The average velocity of light in the substance will therefore be

$$V' = \frac{a+b+a+\beta}{\dfrac{a+\beta}{v}} = v\left(\frac{a+b}{a+\beta}+1\right) \; .$$

If μ is the index of refraction of the medium at rest, since the optical path in the medium is $va+b$,

$$\mu = \frac{va+b}{a+b} \; ;$$

whence $va = \mu(a+b)-b$, which, substituted in the preceding expression, gives

$$V' = \frac{V}{\mu} + \left(1 - \frac{1}{\mu^2} \cdot \frac{b}{a+b}\right) v \; .$$

The quantity in parentheses corresponds to the coefficient of entrainment and coincides the more nearly with Fresnel's coefficient the smaller the value of the ratio a/b.[1]

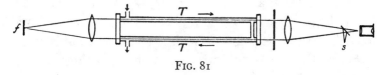

FIG. 81

Notwithstanding these deductions of the Fresnel coefficient from negative experiment as well as theoretical reasoning, this consequence appeared difficult of general acceptance. The result was, however, amply confirmed by an ingenious experiment of Fizeau in which the displacement of interference fringes, due to the combination of two pencils of light traversing a column of liquid in motion in opposite directions as shown schematically in Figure 81, was found to be of the order predicted by theory.

If the interference fringes were to be observed at f, the expected displacement due to the current of water in

[1] If, on the other hand, $b=0$ (molecules in contact), $\mu=\nu$, and $\rho'=1$, whereas Fresnel's $\rho = 1 - 1/\mu^2$.

the tubes TT would be entirely masked by the much larger effects due to the distortions produced by the varying pressures in reversing the current. Fizeau eliminated such disturbing causes entirely by returning the light by a mirror at f back to the source, part being thrown to one side at s by a plane-parallel glass plate.

Fizeau found a displacement corresponding with that which should follow from Fresnel's coefficient, showing conclusively that the light-waves were accelerated by a fraction of the velocity of water and clearly less than the full amount.

In view of the fundamental importance of this subject, and especially in view of another experiment for testing the possibility of detecting a relative motion between the earth and the ether,[1] it was determined to repeat the experiment of Fizeau in such manner as to avoid certain difficulties and uncertainties inherent in his work. Among these the following may be mentioned: first, the form of interference apparatus employed necessitates the use of a fine aperture and consequent feeble intensity of the light; second, close proximity of the two interfering pencils, involving restricted widths of the water columns (otherwise the interfering pencils meet under so large an angle that the interference fringes are too narrow to observe without excessive magnification and still further enfeeblement of the light); third, the difficulty in utilizing a relatively small portion of the area where the velocity is approximately constant; fourth, the uncertainty in the value of this maximum velocity in terms of the mean velocity found by observation.

These objections are obviated in the interferometer

[1] See p. 150.

arranged as in Figure 82. Figure 83 shows the water circuit, which is entirely disconnected from the interferometer (mounted on brick piers), and the arrangement of valves by which the direction of the current is reversed. The mean current was found by measuring the time required to fill a measured volume in the receiving tank and multiplying by the ratio of areas of tank and tube. To

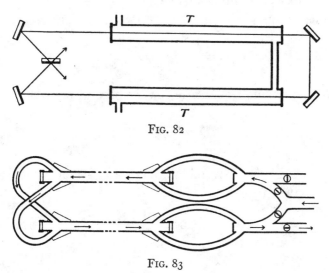

FIG. 82

FIG. 83

find the maximum velocity, a small Pitot tube measured the pressure due to the current, a preliminary calibration giving very consistent readings, showing that the pressures were proportional to the square root of the velocities. This was displaced along the radius of the tube, giving a relation $v = v_m(1 - x^2)^n$, in which v is the velocity at a point x on the radius, v_m is the maximum velocity, and n has the value 0.165. This gives for the mean velocity

$$\bar{v} = \int_0^1 2\pi v x\, dx ,$$

whence

$$v_m = 1.165\bar{v} .$$

With a length L of 3 m and a velocity of 8 m per second, the displacement of the interference fringes was about half the fringe width. In a second series with a length L of 6 m and a velocity of 7 m per second, the displacement was about 0.9 fringe.

The difference in time in the two directions is

$$\partial T = \frac{L}{V/\mu - \rho v} - \frac{L}{V/\mu + \rho v} ,$$

or, omitting small quantities of the second order,

$$\partial T = \frac{2L\mu^2\rho v}{V^2} .$$

This is doubled on reversing the current, whence the displacement in fringes,

$$\Delta = \frac{V\partial T}{\lambda} ,$$

or

$$\Delta = \frac{4L\mu^2\rho v}{\lambda V}$$

whence

$$\rho = \frac{\Delta\lambda V}{4L\mu^2 v} .$$

This gave as the final result

$$\rho = 0.434 \pm .02$$

as against

$$\frac{\mu^2 - 1}{\mu^2} = 0.437 .$$

The problem has since been taken up by Lorentz both theoretically and experimentally. Lorentz finds

$$\rho = \frac{\mu^2 - 1}{\mu^2} - \frac{\lambda}{\mu} \cdot \frac{d\mu}{d\lambda} = 0.451 \, ,$$

which is still within the experimental error. His own experimental results confirm the corrected value.

It appears, therefore, that this remarkable result is amply confirmed both theoretically and experimentally. As a consequence, it has been shown in the two cases considered above that it is impossible to detect a relative motion between the earth and the ether. Lorentz has proved this consequence in a perfectly general manner, at least so far as first-order terms are concerned.

SECOND ORDER EFFECTS

Maxwell was the first to point out that while it must be admitted that there can be no first-order effect which can be brought to light by experiment, this need not necessarily follow for effects depending on the second order. He expressed doubt, however, as to the possibility of detecting such exceedingly small quantities, which may be expected to be of the order of the square of the aberration, i.e., one part in one hundred million.

The length of a light-wave is, however, so small that one hundred million of them make up a distance of 50 m, and if an interferometer be so arranged by repeated reflections from appropriately placed mirrors, the actual dimensions of the apparatus need not be very great in order to obtain a displacement easily measurable if second-order effects are appreciable.

If OA and OB (Fig. 84A) are the two interferometer arms with plane mirrors at A and B which return the two pencils—one reflected, the other transmitted—by the

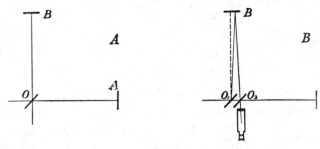

FIG. 84

plane-parallel lightly silvered plate at O, and if OA is in the direction of the earth's motion, then the time required to return to O from A will be longer than from B. In fact, the actual distance traveled by the light to return to the half-silvered plate will be

$$D + \partial_1 + D - \partial_2 \,,$$

where the distance A moves, before the light reaches it, is

$$\partial_1 = \frac{v}{V - v}\, D \,,$$

and the distance O moves, before the light returns, is

$$\partial_2 = \frac{v}{V + v}\, D \,.$$

The total distance neglecting the fourth power of small quantities, is, therefore

$$2D_1 = 2D\left(1 + \frac{v^2}{V^2}\right) \,.$$

But the other path is also affected by the motion; for, in order to meet the dividing surface on return, the actual path will be O_1BO_2 (Fig. $84B$), where

$$O_1O_2 = 2Dv/V \text{ ,}$$

and therefore

$$O_1BO_2 = 2D\sqrt{1 + \frac{v^2}{V^2}} \text{ ,}$$

or, to the same approximation,

$$2D_2 = 2D\left(1 + \frac{v^2}{2V^2}\right) \text{ .}$$

Hence there will be a difference between the two paths expressed in wave-lengths,

$$\Delta = \frac{D}{\lambda} \frac{v^2}{V^2} \text{ .}$$

If now the two directions be interchanged by a rotation through 90°, the total displacement of the interference fringes to be expected will be $\dfrac{2D}{\lambda} \dfrac{v^2}{V^2}$.

In order to minimize the displacements due to external factors, chiefly to distortions produced during rotation, the interferometer was mounted (as shown in Fig. 85) on a block of stone 1.5 m square and .25 m thick resting on an annular wooden ring which floated the whole apparatus on mercury. At each corner of the stone four mirrors *ddee* (Fig. 86), were placed. Near the center of the stone was a plane-parallel lightly silvered glass plate *b*. These were so disposed that light from an Argand burner *a*, passing through a lens, fell on *b*, part

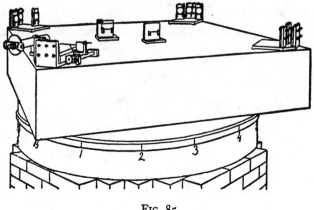

FIG. 85

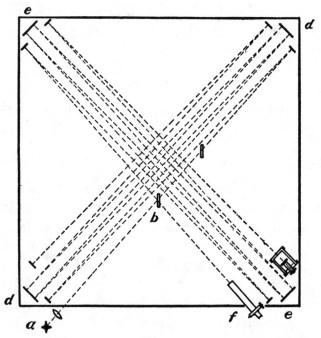

FIG. 86

being transmitted and part reflected, following the paths indicated in the figure. The resulting interference fringes were observed by the telescope f. Both f and a revolve with the stone.

By keeping up a fairly uniform and continuous rotation, observing the position of the central fringe at intervals of one-sixteenth of a revolution, the readings were found to give fairly consistent results, the mean of which

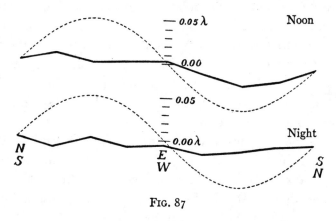

FIG. 87

is represented by Figure 87, in which the dotted curve represents one-eighth of the theoretical displacement.

It must be concluded that the experiment shows no evidence of a displacement greater than 0.01 of a fringe. Considering the motion of the earth in its orbit only, this displacement should be $2D/\lambda \; v^2/V^2$. The distance D was about 11 m, or 2×10^7 waves of yellow light. With $\dfrac{v}{V} = \dfrac{1}{10,000}$, this gives an expected displacement of 0.4 fringe. The actual value is certainly less than one-twentieth of this amount and probably less than one-fortieth.

In what precedes, only the orbital motion of the earth is considered, and it might be possible that this chanced to be just enough to neutralize the motion of the system through space. The experiments have been again taken up by Morley and Miller, using a still larger apparatus, and two series of investigations have resulted in a confirmation of the negative result.

It is obvious from all that precedes that it is impossible to measure any possible motion relative to the ether by observations at the surface of the earth. But it is not impossible that at even moderate distances above sea-level, at the top of a mountain, such an experiment might give a different result.[1] If we allow such a possibility, it would seem logical to admit that at some distance above the earth's surface, let us say 10 miles (or 10,000 miles), such an entrainment would vanish. If, then, an interferometer circuit be set up in a vertical plane parallel to the earth's motion, such a diminution of relative motion should be manifest by a corresponding displacement of the interference fringes.

Thus suppose $v = \overline{V}e^{-.001h}$, which would correspond to a diminution to $1/e$ of the velocity of the earth at a distance from the surface of 1,000 km. For a height h of 10 m the diminution would be only 2×10^{-5}, but for a length of 100 m this would correspond to a displacement of the order of a half-fringe. The experiment was actually tried, but no displacement was observed.

While too much weight should not be given to an

[1] Recent experiments by Dayton C. Miller seem to give a positive result, indicating a small fraction (one thirtieth) of the hypothetical velocity of the galactic system of 300 kilometers per second. Such a result would contradict the principle of relativity. Experiments are now in preparation for a rigorous test.

experiment which is founded on so much pure assumption, it may be worth mentioning that it is in accord with the result of the Fizeau experiment, and also with an experiment of Lodge, who found no displacement in an interferometer circuit placed between two rapidly rotating disks of steel. The objection may, however, be raised that the mass in motion in the last two-mentioned experiments may have been far too small to allow of extrapolation to a body of the size of the earth.

It must be admitted, however, that these experiments are not sufficiently conclusive to justify the hypothesis of an ether which is entrained with the earth in its motion. But then how can the negative results be explained?

CHAPTER XIV*

RELATIVITY

Lorentz and Fitzgerald have proposed a possible solution of the null effect of the Michelson-Morley experiment by assuming a contraction in the material of the support for the interferometer just sufficient to compensate for the theoretical difference in path. Such a hypothesis seems rather artificial, and it of course implies that such contractions are independent of the elastic properties of the material.[1]

The contraction hypothesis of Lorentz leads to the important set of equations known as the "Lorentz transformation," which may be deduced on the basis of an equation expressing the independence of a physical phenomenon on the motion of the observer.

Suppose the phenomenon selected for the purpose to be the propagation of an electromagnetic disturbance. The equation expressing such a wave propagation in the direction of x is

$$\frac{\partial^2 U}{\partial x^2} = \frac{1}{c^2} \cdot \frac{\partial^2 U}{\partial t^2} ,$$

where c is the velocity of light.[2] This equation should remain invariant if new co-ordinates x_1, t_1 are chosen

[1] This consequence was tested by Morley and Miller by substituting a support of wood for that of stone. The result was the same as before.

[2] Laue, *Das Relativitätsprincip*.

* See note, p. v.

referred to co-ordinate axes in uniform motion of translation with respect to the former. Accordingly,

$$\frac{\partial^2 U}{\partial x^2} - \frac{1}{c^2}\frac{\partial^2 U}{\partial t^2} = \frac{\partial^2 U}{\partial x_1^2} - \frac{1}{c^2}\frac{\partial^2 U}{\partial t_1^2} \cdot$$

The simplest relations existing between x_1, t_1 and x, t are

$$x_1 = k(x - vt)$$
$$t_1 = at - \beta x ,$$

where k, a, and β are constants to be determined. Substituting in the preceding equation, we readily find

$$k = \frac{1}{\sqrt{1 - v^2/c^2}} = a ; \quad \beta = \frac{va}{c^2} .$$

From these follow the Lorentz equations,

$$x_1 = \frac{x - vt}{\sqrt{1 - v^2/c^2}}$$

$$t_1 = \frac{t - vx/c^2}{\sqrt{1 - v^2/c^2}} .$$

These results deduced on the basis of the Lorentz contraction in material bodies moving through a fixed ether follow as direct consequences of the (restricted) theory of relativity of Einstein. This postulates (1) that relative motions alone are measurable (and is consequently inconsistent with a medium at rest to which all such motions might be referred); (2) that the velocity of light is a constant independent of the uniform motion of the observer.

The Lorentz contraction is found by expressing the

distance between the co-ordinates of the ends of a meter bar placed parallel to the direction of translation.

$$\Delta x = x' - x'' = \frac{x_1' - x_1''}{\sqrt{1 - v^2/c^2}} ,$$

or

$$\Delta x = k\Delta x_1 ,$$

or

$$\Delta x_1 = \frac{1}{k} \Delta x ;$$

accordingly, the meter bar is shorter when in motion than when at rest in the proportion $\sqrt{1 - v^2/c^2} : 1$. Similar considerations show that a sphere would appear as a flattened ellipsoid. Similarly, it follows that a time interval $\Delta t_1 = \Delta t \sqrt{1 - v^2/c^2}$.

Applying these results to the Michelson-Morley experiment, we find that the time of return of the light which travels in the direction of v,

$$t_1 = \frac{2Dc}{c^2 - v^2} \times \sqrt{1 - v^2/c^2} = \frac{2D}{\sqrt{c^2 - v^2}} ,$$

while in the perpendicular direction

$$t_2 = \frac{2D}{c} \sqrt{1 + v^2/c^2} ,$$

or, neglecting terms of the fourth order,

$$t_2 = \frac{2D}{\sqrt{c^2 - v^2}} = t_1 ,$$

which shows that on this theory there should be no difference in the two times, or in other words, there should be no displacement of the interference fringes.

The absence of first-order effects may be deduced from the formula for the addition of velocities. If v_1 is the velocity in the system $x_1 t_1$, and v the velocity in the system xt, then

$$x_1 = v_1 t_1 , \qquad x = \frac{x_1 + v t_1}{\sqrt{1 - v^2/c^2}} = \frac{(v + v_1) t_1}{\sqrt{1 - v^2/c^2}} ,$$

or

$$x = \frac{v + v_1}{1 + v v_1/c^2} \cdot t ,$$

whence

$$\bar{v} = \frac{v + v_1}{1 + v v_1/c^2} .$$

This result appears somewhat paradoxical in consequence of the difficulty of realizing the actual circumstances under which it applies. For instance, if $v = c$, the velocity of light, then $\bar{v} = c$, a result which may be otherwise stated: There can be no velocity greater than the velocity of light; a statement which must also be true of any of the expressions involving the radical $\sqrt{1 - v^2/c^2}$, which then is meaningless.[1]

Suppose that v_1, the added velocity, is that which occurs in the experiment of Fizeau; in this case the resultant is

$$\bar{v} = \frac{\dfrac{c}{\mu} + v_1}{1 + \dfrac{v_1}{\mu c}} ,$$

[1] On the basis of an ether, which is the seat of all electromagnetic action, and in view of the electronic constitution of matter, all motion must be accompanied by magnetic fields proportional to the velocity which resist the motion; and when the velocity is equal to that of the velocity of the light, the resistance to further increase is enormous if not infinite.

or, neglecting terms of the second order,

$$\bar{v} = \frac{c}{\mu} + \frac{\mu^2 - 1}{\mu^2} v_1 ,$$

which is Fresnel's formula.

Thus the theory gives a consistent explanation of all the negative results obtained in the attempts to detect a relative motion. But besides this, it deduces the increase of mass with velocity which is expressed by

$$m = \frac{m_0}{\sqrt{1 - v^2/c^2}}$$

(m_0 corresponding to the mass when at rest), a result which is confirmed by experiments on high-speed electrons.

The generalized theory of relativity has furnished still more remarkable results. This considers not only uniform but also accelerated motion. In particular, it is based on the impossibility of distinguishing an acceleration from the gravitation or other force which produces it. Three consequences of the theory may be mentioned of which two have been confirmed while the third is still on trial: (1) It gives a correct explanation of the residual motion of forty-three seconds of arc per century of the perihelion of Mercury. (2) It predicts the deviation which a ray of light from a star should experience on passing near a large gravitating body, the sun, namely, 1″.7. On Newton's corpuscular theory this should be only half as great. As a result of the measurements of the photographs of the eclipse of 1921 the number found was much nearer to the prediction of Einstein, and was inversely proportional to the distance from the center of the sun, in further confirmation of the theory. (3) The theory predicts a displacement of the solar spectral lines, and it seems that this prediction is also verified.

Thus the theory of relativity has not only furnished an explanation of known phenomena, but has made it possible to predict and to discover new phenomena, which is one of the most convincing proofs of the value of a theory. It must therefore be accorded a generous acceptance notwithstanding the many consequences which may appear paradoxical in consequence of the difficulty we find in realizing the unusual conditions of high relative velocities.

The existence of an ether appears to be inconsistent with the theory; a fixed ether would imply the possibility of a measurement of "absolute" motion. But without a medium how can the propagation of light-waves be explained? On the electromagnetic theory the velocity of transmission of an electromagnetic disturbance is the square root of the reciprocal of the product of permeability by dielectric constant, which are properties of the medium. How explain the constancy of propagation, the fundamental assumption (at least of the restricted theory), if there be no medium?

It is true that several attempts have been made to overcome this objection; resuscitations of the exploded corpuscular theory, propagation along lines of force, etc.; but these not only raise many more difficulties than they explain, but are totally inadequate to account for the constancy of propagation.

It is to be hoped that the theory may be reconciled with the existence of a medium, either by modifying the theory, or, more probably, by attributing the requisite properties to the ether; for example, allowing changes in its properties (dielectric constant, for instance) due to the presence of a gravitational field.

The result that it is impossible to obtain a first-order effect in the motion of the earth relative to the ether is a conclusion of the theory of a fixed ether (*vide supra*) as well as of the theory of relativity. The following experiment, if it could be carried out, should also give a negative result.

If *A* and *B* are two Fizeau wheels revolving with equal speeds, and a pencil of light pass between the teeth at *A* and also at *B*, then (on the fixed ether theory) if the whole is in motion with the earth in the direction *AB*, a pencil returning in the opposite direction would, at a certain speed of rotation, be intercepted and would not be visible on the other side. The experiment, if practicable, should therefore be capable of measuring such a relative motion. If, in accordance with relativity, this is impossible, it is equivalent to the denial of the possibility of securing exactly equal speeds in the two wheels. This would certainly be true if the wheels were actuated by electromagnetic control, for this would be subject to the same influence as the light-waves themselves.

If the connection were material, by means of an axle of sufficient rigidity, the coincidence could be secured, but here again it may be objected that the intermolecular forces, which determine the elasticity and therefore the velocity of propagation, which are probably electrical, would also be affected in the same way.[1]

[1] In a preliminary effort to utilize acoustic vibrations it was found possible to transmit the vibrations of a large tuning fork through a stretched piano wire over a distance of a mile, with an amplitude of vibra-

Another possibility of measurement of the velocity of light in one direction is based on the observation of the eclipses of Jupiter's satellites. By comparing the results when the direction earth–Jupiter is nearest the direction of the motion of the solar system with the corresponding value when this direction is reversed, the deductions of the Einstein theory might be tested. According to this theory, there should be no difference; while, on the basis of a stationary ether, there should be a difference of the order of two-tenths of a second in the time required for light to cross the earth's orbit.

The following experiment may possibly furnish a test of ether entrainment, although on the basis of a fixed ether as well as in accordance with the (generalized) relativity the result should be the same, namely, a displacement of the interference fringes of the order of one fringe for an area of 1 sq. km.

The experiment consists in observing the position of the fringes in an interferometer circuit of this size, or less, and comparing the position of the fringes with that of a much smaller auxiliary circuit. If these two circuits have images of the source (a minute aperture or a narrow slit) which accurately coincide, then the interference fringes of the two systems will also coincide, if the effect to be expected is zero. But if the expected displacement is that required by both theories, this difference, amounting to 0.5 to 1.0 fringe, can be observed with certainty.

Preliminary trial has shown that it is possible to observe fringes with a triangular circuit of 1,500 m. In

tion in a second fork of the same pitch about half as great as that of the primary. Doubtless the same objection holds in this proposed method, yet it might be of interest to test the matter in some such fashion.

case the expected displacement is observed, it would add a little to the evidence for relativity (or as much for a fixed ether). But if the displacement should turn out to be zero, or even appreciably less than the calculated amount, such a result could hardly be reconciled with any hypothesis save that of an ether which is entrained with the earth in its rotation.

In a series of preliminary trials made at Mount Wilson, the atmospheric disturbances were too great over the circuit of the interfering pencils of light to permit of accurate measurement. It was necessary, therefore, to allow the light to proceed in a partial vacuum in a pipe line something over a mile in length, shown diagrammatically in Figure 88. The experiment was tried under these conditions at Clearing some 10 miles southwest of Chicago. The light-path is indicated by the dotted line $ADEFA$, an auxiliary path $ADCB$ furnishing a comparison set of interference fringes corresponding to an inclosed area, small compared with the former.

The calculated value of the displacement of the fringes on the assumption of a stationary ether as well as in accordance with relativity is readily shown to be[1]

$$\Delta = \frac{4A\omega \sin \phi}{\lambda V},$$

where Δ is the expected displacement in fringe widths, A the area of the light-path, ϕ the latitude, ω the angular velocity of the earth, λ the effective wave-length of the light, and V the velocity of light.

In this latitude, 41°40′, with a pipe line about 2,000 by 1,100 feet and an effective wave-length of 0″57, this

[1] See L. Silberstein, *Journal of the Optical Society*, 5, 291, 1921.

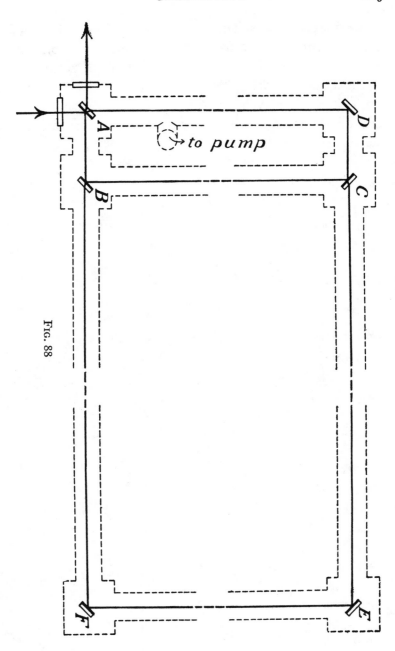

to pump

FIG. 88

gives an expected displacement of the interference fringes of 0.236 fringe. The observed displacement was 0.230, an agreement which is well within the limit of accuracy of the measurement.

This result may be considered as an additional evidence in favor of relativity—or equally as evidence of a stationary ether.

CHAPTER XV

METALLIC COLORS IN BIRDS
AND INSECTS

It is hoped that the results of the investigations recorded in the preceding chapters may be considered as fairly certain contributions to the science of optics. In this chapter on "Metallic Colors" it is only fair to state that the subject is still under discussion, and the following presentation is chiefly concerned with evidence in favor of the hypothesis of a thin surface film of high reflecting power.

There have been numerous attempts to explain the beautiful colors exhibited by the plumage of humming-birds, peacocks, and pigeons, and many butterflies and beetles,[1] the chief characteristics of which are: (1) an unusually high intensity of the reflected light, especially for some one color at normal incidence; (2) A change in color with the angle of incidence usually (but not always) toward the violet. These effects are clearly evident to the unaided eye. The following, requiring instrumental appliances, may be added: (3) The distribution of the reflected light in the spectrum is not what should be expected from interference. This should show either a channeled spectrum if the layer space is more than half a light-wave, or else a single narrow bright band, neither of

[1] The occurrence of metallic colors in the vegetable kingdom is very rare.

which corresponds to the actual distribution which, in almost all the cases examined, shows a very broad band covering half or more of the visible spectrum. (4) The law of change of wave-length of the light-maximum with the angle of incidence differs decisively from that which follows as a result of interference. (5) The reflected light is always elliptically polarized, a quality very characteristic of metals and of other intensely absorbing substances such as the aniline dyes. (6) Direct evidence of a single effective layer generally of thickness small compared with a light-wave.

Two theories have been proposed to account for these effects. The first is due to Lord Rayleigh, and attributes the entire effect to the result of repeated reflections from approximately equidistant layers of alternating optical properties such as are typified in the crystals of potassium chlorate. According to the second theory, the result is attributed chiefly to a very thin single layer of high absorptive power such as is shown by the aniline dyes, and in some cases more closely resembling the qualities of actual metals.

The reflecting power of a specimen of the wing-case of a beetle (*Plusiotis resplendens*) was found to be equal to that of "tin" foil, and its appearance is scarcely to be distinguished from brass.

To account for so high a reflecting power on the interference hypothesis would require a large number of reflecting layers of which there is no evidence in the specimen as shown by polishing away a portion of the wingcase as illustrated in Plate 1*A*[1].

Similar effects are shown in Plates 1*B* and 1*C*, in which a portion of the wing-case of a brilliant-green beetle was

[1]Plate 1 (*A–C*) is reproduced in color on the inside front cover.

PLATE I

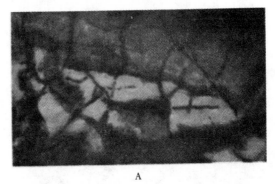

A

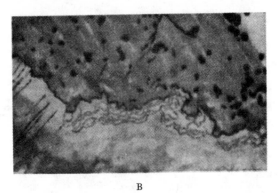

B

C

removed. In these specimens, although layers are clearly indicated in the portion outside of the characteristic green, there is no evidence that these contribute in any way to produce the green color, since there is a complete absence of color gradation.

In all three cases the demarcation between the color films is so abrupt that the "width" of the edge is below the limit of resolution of the microscope.

Taking this as about half a micron, the upper limit of the thickness of the film may be calculated. If a is the width of the elliptical hole in the wing-case, and R the shorter radius of curvature, the thickness of the layer is given by

$$t = h\frac{a}{2R}$$

in which h is the limit of resolution. In the present case, $h = 0^{\mu}5$, $a = 0.7$ mm, $R = 8$ mm, whence $t = 0^{\mu}022$; or the thickness of the film is less than a twentieth of a light-wave.

The reflecting power of a "silver-dotted" moth is of the order of 30 to 50 per cent. The surface of the scale is made up of equidistant longitudinal ridges (about a thousand to the millimeter) with very fine transverse markings at about half this spacing. The reflection occurs at the individual ridges, and explanation of the high reflecting power by successive layers seems wholly debarred.

The difference between the law of color change with angle of incidence as calculated on the basis of succession of equidistant layers and the observed values of the maximum wave-length is well illustrated in Figure 84.

The dotted curves represent the value of λ, the wavelength, for maximum intensity as calculated from

$$\lambda = \lambda_0 \sqrt{1 - \frac{\sin^2 i}{\mu^2}}$$

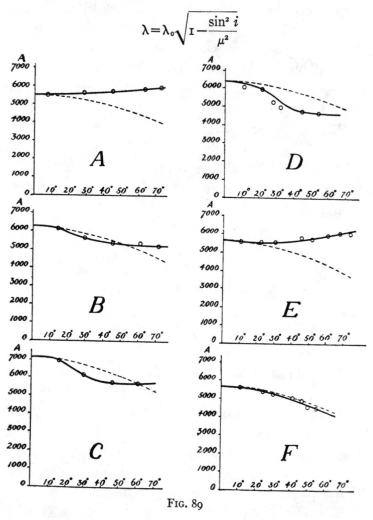

FIG. 89

where i is the angle of incidence and μ the index of refraction (assumed 1.5 in the calculation) while the full curve represents the results of observation.

It will be noted that in every case excepting *F*, there is a marked difference between the observed and the calculated values. This is most evident in the specimen already mentioned as the "brass beetle" (*Plusiotis resplen-*

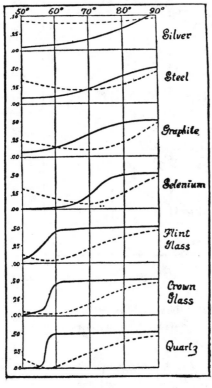

FIG. 90

dens) in Figure 89*A* in which the change of color is toward the red end of the spectrum with oblique incidence.

Figure 89*B* is from the brilliant-orange throat feather of a Brazilian humming-bird. Figure 89*C* represents the curves found in the coppery wing-case of a beetle.

Figure 89*D* shows the results for the "diamond" beetle. Figure 89*E* illustrates the wide divergence in the case of a green-winged butterfly.

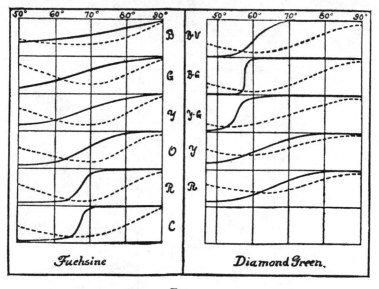

FIG. 91

Figure 89*F*, in contrast with these, shows the close correspondence between the theoretical and the observed values for the opal.

One of the most important characteristics of metallic reflection is the difference of phase, Δ, introduced between the components polarized in the plane of incidence, and perpendicular thereto. This is a function of the angle of incidence, as illustrated in the graphs reproduced in Figure 90.[1]

It will be noted that transparent substances are char-

[1] Full curves; the dotted curves represent the ratio of the two components after reflection.

acterized by a very steep portion of the curve in the vicinity of the polarizing angle, whereas the steepness of curves for highly absorbing substances, notably metals, is much less.

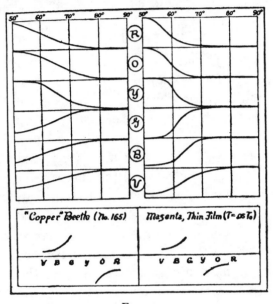

FIG. 92

These conclusions are justified also on theoretical grounds which give as a result

$$\frac{d\Delta}{di} = \frac{\sin I \ (2+\tan^2 I)}{K}$$

in which I is the angle of incidence corresponding to $\Delta = \frac{\pi}{2}$, and K the coefficient of absorption. The steepness of the curve at this point is a measure of the transparency —or reciprocally, the absorption is inversely proportional to $\dfrac{d\Delta}{di}$. In the case of substances such as the aniline dyes, the absorption is intense for some colors, while for others they are almost perfectly transparent. For such we should expect a system of curves showing corresponding variations in $\dfrac{d\Delta}{di}$. This anticipation is fully justified as shown in the graphs for fuchsine and diamond green (Fig. 89): the former showing "metallic" reflection except for red light for which it is as transparent as glass, and the latter showing intense opacity except for the blue green.

If, however, the reflecting film is very thin (of the order of 1/10 of a light-wave) the curves are characterized by an inversion as shown in the right-hand portion of Figure 91 for magenta.

The curves on the left are those observed in the case of a "copper" beetle, and the resemblance is so close that there can be no doubt that in both cases the effective cause of the "metallic" color is a very thin layer with a large value of K except for yellow-green light which is freely transmitted.

In every case thus far examined of "metallic" colors of feathers, butterfly scales, or beetle wing-cases the curves are of this character.

INDEX

Aberration, 139
Accuracy, limit of, 78
Airy, 65, 139
Anderson, J. A., 116
Angles, testing of, 74, 134
Arago, 124

Babinet's theorem, 68
Biprism, Fresnel, 16
Bradley, 121
Bureau of Standards, 54

Cadmium radiations, wave-lengths of, 52
Clearing, experiment at, 164
Coast and Geodetic Survey, 136
Cornu, 124
Coronas, 69
Corpuscular theory, 1

Diffraction, 55; at circular aperture, 63; at rectangular aperture, 61; at two similar rectangular apertures, 65; gratings, 86; gratings, errors of, 92; gratings, ruling of, 99; screen, 59
Dispersion: of grating, 87; of prism, 88
Double star, angular separation of components by interferometer, 112, 116

Echelon grating, 104
Einstein, 157
Eisenlohr, 143
Errors of diffraction gratings, 92

Fabry-Perot: "distance pieces," 52; apparatus, 108, limit of accuracy in, 83
Films, colors of thin, 12

Fizeau, 4, 120, 122, 145
Foucault, 4, 120, 125, 127, 129, 130
Frauenhofer, 65
Fresnel biprism, 16; coefficient, 139; mirrors, 15

Ghosts, 97
Gibbs, 132
Grating spectra, brightness of, 89
Gratings: diffraction, 86; echelon, 163; errors of, 92; phase, 90; ruling of, 99
Gravity meter, 85
Group velocity, 131

Hewitt, 80
Huyghens' principle, 3

Imperfections in optical instruments, 69
Interference, 1, 10; apparatus, forms of, 17, 100; applications of, to astronomy 111; conditions for, 17; effect of "seeing" on, 115
Interferometer, 20, 26; analogy with lenses and gratings, 21; applications of, 20, 22, 100, 111; theory of, 28

Jamin refractometer, 19

Kater pendulum, 53

Limit of accuracy in optical measurements, 78; echelon, 109; effect of "seeing" on, 80; Fabry-Perot etalon, 109; rectangular aperture, 63, 87; resolution, circular aperture, 63; tolerance, 71
Listing, 124
Littrow mounting, 102

Lorentz, 49
Lorentz-Fitzgerald contraction hypothesis, 156
Lorentz transformation, 156

Maxwell, 149
Metallic colors, 167
Meter, measurement of standard, 46
Michelson-Morley experiment, 150
Microscope, testing of, 81
Miller, 154
Morley, 21, 150
Moulton, 114

Newcomb, 129
Newton's rings, 12

Opacity screen, 59

Pease, 119
Perot, Fabry and, 52, 83, 108
Perrotin, 124
Phase: change of, on reflection, 13; factor, 59; grating, 90
Plane parallel plates, 74

Rayleigh, 43, 63, 70, 71, 130, 168
Relativity, 4, 156

Resolving power, 62; of grating, 87, 90; of prism, 88
Roemer, 121

Schwerd, 65
Sight tube, limit of accuracy in, 82
Silberstein, 164
Spherical surfaces, testing of, 75
Standard decimeter, 54
Standard meter, 51

Testing: of angles, 74, 134; of optical surfaces, 73
Thompson, Elihu, 80
Tolerance, limit of, 71

Undulatory theory, 1

Velocity, group, 131
Velocity of light, 120; effect of motion of medium on, 139; in water, 2, 124, 130
Visibility curves, 34; of fringes, 31

Wave-length, 14, 15
Wave-trains, 7
Wheatstone, 14

Young's experiment, 13

A CATALOG OF SELECTED
DOVER BOOKS
IN SCIENCE AND MATHEMATICS

A CATALOG OF SELECTED
DOVER BOOKS
IN SCIENCE AND MATHEMATICS

QUALITATIVE THEORY OF DIFFERENTIAL EQUATIONS, V.V. Nemytskii and V.V. Stepanov. Classic graduate-level text by two prominent Soviet mathematicians covers classical differential equations as well as topological dynamics and ergodic theory. Bibliographies. 523pp. 5⅜ × 8½.   65954-2 Pa. $10.95

MATRICES AND LINEAR ALGEBRA, Hans Schneider and George Phillip Barker. Basic textbook covers theory of matrices and its applications to systems of linear equations and related topics such as determinants, eigenvalues and differential equations. Numerous exercises. 432pp. 5⅜ × 8½.   66014-1 Pa. $10.95

QUANTUM THEORY, David Bohm. This advanced undergraduate-level text presents the quantum theory in terms of qualitative and imaginative concepts, followed by specific applications worked out in mathematical detail. Preface. Index. 655pp. 5⅜ × 8½.   65969-0 Pa. $13.95

ATOMIC PHYSICS (8th edition), Max Born. Nobel laureate's lucid treatment of kinetic theory of gases, elementary particles, nuclear atom, wave-corpuscles, atomic structure and spectral lines, much more. Over 40 appendices, bibliography. 495pp. 5⅜ × 8½.   65984-4 Pa. $12.95

ELECTRONIC STRUCTURE AND THE PROPERTIES OF SOLIDS: The Physics of the Chemical Bond, Walter A. Harrison. Innovative text offers basic understanding of the electronic structure of covalent and ionic solids, simple metals, transition metals and their compounds. Problems. 1980 edition. 582pp. 6⅛ × 9¼.   66021-4 Pa. $15.95

BOUNDARY VALUE PROBLEMS OF HEAT CONDUCTION, M. Necati Özisik. Systematic, comprehensive treatment of modern mathematical methods of solving problems in heat conduction and diffusion. Numerous examples and problems. Selected references. Appendices. 505pp. 5⅜ × 8½.   65990-9 Pa. $12.95

A SHORT HISTORY OF CHEMISTRY (3rd edition), J.R. Partington. Classic exposition explores origins of chemistry, alchemy, early medical chemistry, nature of atmosphere, theory of valency, laws and structure of atomic theory, much more. 428pp. 5⅜ × 8½. (Available in U.S. only)   65977-1 Pa. $10.95

A HISTORY OF ASTRONOMY, A. Pannekoek. Well-balanced, carefully reasoned study covers such topics as Ptolemaic theory, work of Copernicus, Kepler, Newton, Eddington's work on stars, much more. Illustrated. References. 521pp. 5⅜ × 8½.   65994-1 Pa. $12.95

PRINCIPLES OF METEOROLOGICAL ANALYSIS, Walter J. Saucier. Highly respected, abundantly illustrated classic reviews atmospheric variables, hydrostatics, static stability, various analyses (scalar, cross-section, isobaric, isentropic, more). For intermediate meteorology students. 454pp. 6⅛ × 9¼.   65979-8 Pa. $14.95

RELATIVITY, THERMODYNAMICS AND COSMOLOGY, Richard C. Tolman. Landmark study extends thermodynamics to special, general relativity; also applications of relativistic mechanics, thermodynamics to cosmological models. 501pp. 5⅜ × 8½. 65383-8 Pa. $12.95

APPLIED ANALYSIS, Cornelius Lanczos. Classic work on analysis and design of finite processes for approximating solution of analytical problems. Algebraic equations, matrices, harmonic analysis, quadrature methods, much more. 559pp. 5⅜ × 8½. 65656-X Pa. $13.95

SPECIAL RELATIVITY FOR PHYSICISTS, G. Stephenson and C.W. Kilmister. Concise elegant account for nonspecialists. Lorentz transformation, optical and dynamical applications, more. Bibliography. 108pp. 5⅜ × 8½. 65519-9 Pa. $4.95

INTRODUCTION TO ANALYSIS, Maxwell Rosenlicht. Unusually clear, accessible coverage of set theory, real number system, metric spaces, continuous functions, Riemann integration, multiple integrals, more. Wide range of problems. Undergraduate level. Bibliography. 254pp. 5⅜ × 8½. 65038-3 Pa. $7.95

INTRODUCTION TO QUANTUM MECHANICS With Applications to Chemistry, Linus Pauling & E. Bright Wilson, Jr. Classic undergraduate text by Nobel Prize winner applies quantum mechanics to chemical and physical problems. Numerous tables and figures enhance the text. Chapter bibliographies. Appendices. Index. 468pp. 5⅜ × 8½. 64871-0 Pa. $11.95

ASYMPTOTIC EXPANSIONS OF INTEGRALS, Norman Bleistein & Richard A. Handelsman. Best introduction to important field with applications in a variety of scientific disciplines. New preface. Problems. Diagrams. Tables. Bibliography. Index. 448pp. 5⅜ × 8½. 65082-0 Pa. $12.95

MATHEMATICS APPLIED TO CONTINUUM MECHANICS, Lee A. Segel. Analyzes models of fluid flow and solid deformation. For upper-level math, science and engineering students. 608pp. 5⅜ × 8½. 65369-2 Pa. $13.95

ELEMENTS OF REAL ANALYSIS, David A. Sprecher. Classic text covers fundamental concepts, real number system, point sets, functions of a real variable, Fourier series, much more. Over 500 exercises. 352pp. 5⅜ × 8½. 65385-4 Pa. $10.95

PHYSICAL PRINCIPLES OF THE QUANTUM THEORY, Werner Heisenberg. Nobel Laureate discusses quantum theory, uncertainty, wave mechanics, work of Dirac, Schroedinger, Compton, Wilson, Einstein, etc. 184pp. 5⅜ × 8½.
60113-7 Pa. $5.95

INTRODUCTORY REAL ANALYSIS, A.N. Kolmogorov, S.V. Fomin. Translated by Richard A. Silverman. Self-contained, evenly paced introduction to real and functional analysis. Some 350 problems. 403pp. 5⅜ × 8½. 61226-0 Pa. $9.95

PROBLEMS AND SOLUTIONS IN QUANTUM CHEMISTRY AND PHYSICS, Charles S. Johnson, Jr. and Lee G. Pedersen. Unusually varied problems, detailed solutions in coverage of quantum mechanics, wave mechanics, angular momentum, molecular spectroscopy, scattering theory, more. 280 problems plus 139 supplementary exercises. 430pp. 6½ × 9¼. 65236-X Pa. $12.95

CATALOG OF DOVER BOOKS

ASYMPTOTIC METHODS IN ANALYSIS, N.G. de Bruijn. An inexpensive, comprehensive guide to asymptotic methods—the pioneering work that teaches by explaining worked examples in detail. Index. 224pp. 5⅜ × 8½. 64221-6 Pa. $6.95

OPTICAL RESONANCE AND TWO-LEVEL ATOMS, L. Allen and J.H. Eberly. Clear, comprehensive introduction to basic principles behind all quantum optical resonance phenomena. 53 illustrations. Preface. Index. 256pp. 5⅜ × 8½. 65533-4 Pa. $7.95

COMPLEX VARIABLES, Francis J. Flanigan. Unusual approach, delaying complex algebra till harmonic functions have been analyzed from real variable viewpoint. Includes problems with answers. 364pp. 5⅜ × 8½. 61388-7 Pa. $8.95

ATOMIC SPECTRA AND ATOMIC STRUCTURE, Gerhard Herzberg. One of best introductions; especially for specialist in other fields. Treatment is physical rather than mathematical. 80 illustrations. 257pp. 5⅜ × 8½. 60115-3 Pa. $6.95

APPLIED COMPLEX VARIABLES, John W. Dettman. Step-by-step coverage of fundamentals of analytic function theory—plus lucid exposition of five important applications: Potential Theory; Ordinary Differential Equations; Fourier Transforms; Laplace Transforms; Asymptotic Expansions. 66 figures. Exercises at chapter ends. 512pp. 5⅜ × 8½. 64670-X Pa. $11.95

ULTRASONIC ABSORPTION: An Introduction to the Theory of Sound Absorption and Dispersion in Gases, Liquids and Solids, A.B. Bhatia. Standard reference in the field provides a clear, systematically organized introductory review of fundamental concepts for advanced graduate students, research workers. Numerous diagrams. Bibliography. 440pp. 5⅜ × 8½. 64917-2 Pa. $11.95

UNBOUNDED LINEAR OPERATORS: Theory and Applications, Seymour Goldberg. Classic presents systematic treatment of the theory of unbounded linear operators in normed linear spaces with applications to differential equations. Bibliography. 199pp. 5⅜ × 8½. 64830-3 Pa. $7.95

LIGHT SCATTERING BY SMALL PARTICLES, H.C. van de Hulst. Comprehensive treatment including full range of useful approximation methods for researchers in chemistry, meteorology and astronomy. 44 illustrations. 470pp. 5⅜ × 8½. 64228-3 Pa. $11.95

CONFORMAL MAPPING ON RIEMANN SURFACES, Harvey Cohn. Lucid, insightful book presents ideal coverage of subject. 334 exercises make book perfect for self-study. 55 figures. 352pp. 5⅜ × 8¼. 64025-6 Pa. $9.95

OPTICKS, Sir Isaac Newton. Newton's own experiments with spectroscopy, colors, lenses, reflection, refraction, etc., in language the layman can follow. Foreword by Albert Einstein. 532pp. 5⅜ × 8½. 60205-2 Pa. $9.95

GENERALIZED INTEGRAL TRANSFORMATIONS, A.H. Zemanian. Graduate-level study of recent generalizations of the Laplace, Mellin, Hankel, K. Weierstrass, convolution and other simple transformations. Bibliography. 320pp. 5⅜ × 8½. 65375-7 Pa. $8.95

THE ELECTROMAGNETIC FIELD, Albert Shadowitz. Comprehensive undergraduate text covers basics of electric and magnetic fields, builds up to electromagnetic theory. Also related topics, including relativity. Over 900 problems. 768pp. 5⅜ × 8¼. 65660-8 Pa. $18.95

FOURIER SERIES, Georgi P. Tolstov. Translated by Richard A. Silverman. A valuable addition to the literature on the subject, moving clearly from subject to subject and theorem to theorem. 107 problems, answers. 336pp. 5⅜ × 8½. 63317-9 Pa. $8.95

THEORY OF ELECTROMAGNETIC WAVE PROPAGATION, Charles Herach Papas. Graduate-level study discusses the Maxwell field equations, radiation from wire antennas, the Doppler effect and more. xiii + 244pp. 5⅜ × 8½. 65678-0 Pa. $6.95

DISTRIBUTION THEORY AND TRANSFORM ANALYSIS: An Introduction to Generalized Functions, with Applications, A.H. Zemanian. Provides basics of distribution theory, describes generalized Fourier and Laplace transformations. Numerous problems. 384pp. 5⅜ × 8½. 65479-6 Pa. $9.95

THE PHYSICS OF WAVES, William C. Elmore and Mark A. Heald. Unique overview of classical wave theory. Acoustics, optics, electromagnetic radiation, more. Ideal as classroom text or for self-study. Problems. 477pp. 5⅜ × 8½. 64926-1 Pa. $12.95

CALCULUS OF VARIATIONS WITH APPLICATIONS, George M. Ewing. Applications-oriented introduction to variational theory develops insight and promotes understanding of specialized books, research papers. Suitable for advanced undergraduate/graduate students as primary, supplementary text. 352pp. 5⅜ × 8½. 64856-7 Pa. $8.95

A TREATISE ON ELECTRICITY AND MAGNETISM, James Clerk Maxwell. Important foundation work of modern physics. Brings to final form Maxwell's theory of electromagnetism and rigorously derives his general equations of field theory. 1,084pp. 5⅜ × 8½. 60636-8, 60637-6 Pa., Two-vol. set $21.90

AN INTRODUCTION TO THE CALCULUS OF VARIATIONS, Charles Fox. Graduate-level text covers variations of an integral, isoperimetrical problems, least action, special relativity, approximations, more. References. 279pp. 5⅜ × 8½. 65499-0 Pa. $7.95

HYDRODYNAMIC AND HYDROMAGNETIC STABILITY, S. Chandrasekhar. Lucid examination of the Rayleigh-Benard problem; clear coverage of the theory of instabilities causing convection. 704pp. 5⅜ × 8¼. 64071-X Pa. $14.95

CALCULUS OF VARIATIONS, Robert Weinstock. Basic introduction covering isoperimetric problems, theory of elasticity, quantum mechanics, electrostatics, etc. Exercises throughout. 326pp. 5⅜ × 8½. 63069-2 Pa. $8.95

DYNAMICS OF FLUIDS IN POROUS MEDIA, Jacob Bear. For advanced students of ground water hydrology, soil mechanics and physics, drainage and irrigation engineering and more. 335 illustrations. Exercises, with answers. 784pp. 6⅛ × 9¼. 65675-6 Pa. $19.95

TENSOR CALCULUS, J.L. Synge and A. Schild. Widely used introductory text covers spaces and tensors, basic operations in Riemannian space, non-Riemannian spaces, etc. 324pp. 5⅜ × 8¼. 63612-7 Pa. $8.95

A CONCISE HISTORY OF MATHEMATICS, Dirk J. Struik. The best brief history of mathematics. Stresses origins and covers every major figure from ancient Near East to 19th century. 41 illustrations. 195pp. 5⅜ × 8½. 60255-9 Pa. $7.95

A SHORT ACCOUNT OF THE HISTORY OF MATHEMATICS, W.W. Rouse Ball. One of clearest, most authoritative surveys from the Egyptians and Phoenicians through 19th-century figures such as Grassman, Galois, Riemann. Fourth edition. 522pp. 5⅜ × 8½. 20630-0 Pa. $10.95

HISTORY OF MATHEMATICS, David E. Smith. Nontechnical survey from ancient Greece and Orient to late 19th century; evolution of arithmetic, geometry, trigonometry, calculating devices, algebra, the calculus. 362 illustrations. 1,355pp. 5⅜ × 8½. 20429-4, 20430-8 Pa., Two-vol. set $23.90

THE GEOMETRY OF RENÉ DESCARTES, René Descartes. The great work founded analytical geometry. Original French text, Descartes' own diagrams, together with definitive Smith-Latham translation. 244pp. 5⅜ × 8½. 60068-8 Pa. $7.95

THE ORIGINS OF THE INFINITESIMAL CALCULUS, Margaret E. Baron. Only fully detailed and documented account of crucial discipline: origins; development by Galileo, Kepler, Cavalieri; contributions of Newton, Leibniz, more. 304pp. 5⅜ × 8½. (Available in U.S. and Canada only) 65371-4 Pa. $9.95

THE HISTORY OF THE CALCULUS AND ITS CONCEPTUAL DEVELOPMENT, Carl B. Boyer. Origins in antiquity, medieval contributions, work of Newton, Leibniz, rigorous formulation. Treatment is verbal. 346pp. 5⅜ × 8½. 60509-4 Pa. $8.95

THE THIRTEEN BOOKS OF EUCLID'S ELEMENTS, translated with introduction and commentary by Sir Thomas L. Heath. Definitive edition. Textual and linguistic notes, mathematical analysis. 2,500 years of critical commentary. Not abridged. 1,414pp. 5⅜ × 8½. 60088-2, 60089-0, 60090-4 Pa., Three-vol. set $29.85

GAMES AND DECISIONS: Introduction and Critical Survey, R. Duncan Luce and Howard Raiffa. Superb nontechnical introduction to game theory, primarily applied to social sciences. Utility theory, zero-sum games, n-person games, decision-making, much more. Bibliography. 509pp. 5⅜ × 8½. 65943-7 Pa. $12.95

THE HISTORICAL ROOTS OF ELEMENTARY MATHEMATICS, Lucas N.H. Bunt, Phillip S. Jones, and Jack D. Bedient. Fundamental underpinnings of modern arithmetic, algebra, geometry and number systems derived from ancient civilizations. 320pp. 5⅜ × 8½. 25563-8 Pa. $8.95

CALCULUS REFRESHER FOR TECHNICAL PEOPLE, A. Albert Klaf. Covers important aspects of integral and differential calculus via 756 questions. 566 problems, most answered. 431pp. 5⅜ × 8½. 20370-0 Pa. $8.95

CATALOG OF DOVER BOOKS

CHALLENGING MATHEMATICAL PROBLEMS WITH ELEMENTARY SOLUTIONS, A.M. Yaglom and I.M. Yaglom. Over 170 challenging problems on probability theory, combinatorial analysis, points and lines, topology, convex polygons, many other topics. Solutions. Total of 445pp. 5⅜ × 8½. Two-vol. set.

Vol. I 65536-9 Pa. $7.95
Vol. II 65537-7 Pa. $6.95

FIFTY CHALLENGING PROBLEMS IN PROBABILITY WITH SOLUTIONS, Frederick Mosteller. Remarkable puzzlers, graded in difficulty, illustrate elementary and advanced aspects of probability. Detailed solutions. 88pp. 5⅜ × 8½.
65355-2 Pa. $4.95

EXPERIMENTS IN TOPOLOGY, Stephen Barr. Classic, lively explanation of one of the byways of mathematics. Klein bottles, Moebius strips, projective planes, map coloring, problem of the Koenigsberg bridges, much more, described with clarity and wit. 43 figures. 210pp. 5⅜ × 8½.
25933-1 Pa. $5.95

RELATIVITY IN ILLUSTRATIONS, Jacob T. Schwartz. Clear nontechnical treatment makes relativity more accessible than ever before. Over 60 drawings illustrate concepts more clearly than text alone. Only high school geometry needed. Bibliography. 128pp. 6⅛ × 9¼.
25965-X Pa. $6.95

AN INTRODUCTION TO ORDINARY DIFFERENTIAL EQUATIONS, Earl A. Coddington. A thorough and systematic first course in elementary differential equations for undergraduates in mathematics and science, with many exercises and problems (with answers). Index. 304pp. 5⅜ × 8½.
65942-9 Pa. $8.95

FOURIER SERIES AND ORTHOGONAL FUNCTIONS, Harry F. Davis. An incisive text combining theory and practical example to introduce Fourier series, orthogonal functions and applications of the Fourier method to boundary-value problems. 570 exercises. Answers and notes. 416pp. 5⅜ × 8½.
65973-9 Pa. $9.95

THE THEORY OF BRANCHING PROCESSES, Theodore E. Harris. First systematic, comprehensive treatment of branching (i.e. multiplicative) processes and their applications. Galton-Watson model, Markov branching processes, electron-photon cascade, many other topics. Rigorous proofs. Bibliography. 240pp. 5⅜ × 8½.
65952-6 Pa. $6.95

AN INTRODUCTION TO ALGEBRAIC STRUCTURES, Joseph Landin. Superb self-contained text covers "abstract algebra": sets and numbers, theory of groups, theory of rings, much more. Numerous well-chosen examples, exercises. 247pp. 5⅜ × 8½.
65940-2 Pa. $7.95
